# Hidden Symbolism of
# ALCHEMY and the OCCULT ARTS

# Hidden Symbolism of
# ALCHEMY and the OCCULT ARTS

(formerly titled: *Problems of Mysticism and Its Symbolism*)

# by Dr. Herbert
# Silberer

Translated by SMITH ELY JELLIFFE, M.D., Ph.D.

DOVER PUBLICATIONS, INC.

NEW YORK

This Dover edition, first published in 1971, is an
unabridged and unaltered republication of the work
originally published by Moffat, Yard and Company,
New York, in 1917 under the title *Problems of
Mysticism and its Symbolism*.

*International Standard Book Number: 0-486-20972-5*
*Library of Congress Catalog Card Number: 74-176356*

Manufactured in the United States of America
Dover Publications, Inc.
180 Varick Street
New York, N. Y. 10014

# TRANSLATOR'S PREFACE

Prominent among the stones of a fireplace in my country den, one large rounded giant stands out. It was bourne by the glacial streams from a more northern resting place and is marked by a fossil of a mollusk that inhabited northern seas many million years ago. Yet in spite of the eons of time that have passed it can be compared with specimens of mollusks that live to-day. Down through the countless centuries the living stream has carved its structural habitations in much the same form. The science of Paleontology has collected this history and has attempted a reconstruction of life from its beginnings.

The same principle here illustrated is true for the thought-life of mankind. The forms in which it has been preserved however are not so evident. The structuralizations are not so definite. If they were, evolution would not have been possible for the living stream of energy which is utilized by mind-stuff cannot be confined if it would advance to more complex integrations. Hence the products of mind in evolution are more plastic — more subtle and more changing. They are to be found in the myths and the folk-lore of ancient peoples, the poetry, dramatic art, and the language of later races. From age to age however the strivings continue the same. The living vessels must continue and the products express the most fundamental strivings, in varying though related forms.

We thus arrive at a science which may be called paleo-psychology. Its fossils are the thought-forms throughout the ages, and such a science seeks to show fundamental likenesses behind the more superficial dissimilarities.

The present work is a contribution to such a science in that it shows the essential relationships of what is found in the unconscious of present day mankind to many forms of thinking of the middle ages. These same trends are present to-day in all of us though hidden behind a different set of structural terms, utilizing different mechanisms for energy expression.

The unceasing complexity of life's accumulations has created a great principle for energy expression — it is termed sublimation — and in popular parlance represents the spiritual striving of mankind towards the perfecting of a relation with the world of reality — the environment — which shall mean human happiness in its truest sense. One of the products of this sublimation tendency is called Mysticism. This work would seek to aid us to an understanding of this manifestation of human conduct as expressed in concrete or contemplated action through thought. It does so by the comparative method, and it is for this reason I have been led to present it to an English reading public.

Much of the strange and outre, as well as the commonplace, in human activity conceals energy transformations of inestimable value in the work of sublimation. The race would go mad without it. It sometimes does even with it, a sign that sublimation is still imperfect and that the race is far from being spiritually well. A comprehension of the principles here involved would further the spread of sympathy for all forms of thinking and tend to further spiritual health in such mutual comprehension of the needs of others and of the forms taken by sublimation processes.

For the actual work of translation, I wish to express my obligations to friends Wilfred Lay, and Leo Stein. Without their generous and gifted assistance I would not have been able to accomplish the task.

SMITH ELY JELLIFFE, M.D.

NEW YORK, Oct. 27, 1917.

# CONTENTS

# PART I
## THE PARABLE

## SECTION I

# THE PARABLE

In an old book I discovered an extraordinary narrative entitled Parabola. I take it as the starting point of my observations because it affords a welcome guide. In the endeavor to understand the parable and get a psychological insight into it, we are led on to journey through these very realms of fancy, into which I should like to conduct the reader. At the end of our journey we shall have acquired, with the understanding of the first example, the knowledge of certain psychical laws.

I shall, then, without further prelude introduce the example, and purposely avoid at the outset mentioning the title of the old book so that the reader may be in a position to allow the narrative to affect him without any preconceived ideas. Explanatory interpolations in the text, which come from me, I distinguish with square brackets.

[1]. As once I strolled in a fair forest, young and green, and contemplated the painfulness of this life, and lamented how through the dire fall of our first parents we inherited

such misery and distress, I chanced, while thinking these thoughts, to depart from the usual path, and found myself, I know not how, on a narrow foot path that was rough, untrodden and impassable, and overgrown with so much underbrush and bushes that it was easy to see it was very little used. Therefore I was dismayed and would gladly have gone back, but it was not in my power to do so, since a strong wind so powerfully blew me on, that I could rather take ten steps in advance than one backward.

[2]. Therefore I had to go on and not mind the rough walking.

[3]. After I had advanced a good while I came finally to a lovely meadow hedged about with a round circle of fruit bearing trees, and called by the dwellers *Pratum felicitatis* [the meadow of felicity]. I was in the midst of a company of old men with beards as gray as ice, except for one who was quite a young man with a pointed black beard. Also there was among them one whose name was well known to me, but his visage I could not yet see, who was still younger, and they debated on all kinds of subjects, particularly about a great and lofty mystery, hidden in Nature, which God kept concealed from the great world, and revealed to only a few who loved him.

[4]. I listened long and their discourse pleased me well, only some would break forth from restraint, not touching upon the matter or work, but what touched upon the parables, similitudes and other parerga, in which they followed the poetic fancies of Aristotle, Pliny and others which the one had copied from the other. So I could contain myself no longer and mixed in my own mustard, [put in my own word], refuted such trivial things from experience, and the majority sided with me, examined me in their faculty and made it quite hot for me. However the foundation of my knowledge was so good, that I passed with all honors, where-

upon they all were amazed, unanimously included and admitted me in their collegium, of which I was heartily glad.

[5]. But they said I could not be a real colleague till I learned to know their lion, and became thoroughly acquainted with his powers and abilities. For that purpose I should use diligence so as to subdue him. I was quite confident in myself and promised them I would do my best. For their company pleased me so well that I would not have parted from them for a great deal.

[6]. They led me to the lion and described him very carefully, but what I should undertake with him none could tell me. Some of them indeed hinted, but very darkly, so that the (Der Tausende) Thousandth one could not have understood him. But when I should first succeed in subduing him and should have assured myself against his sharp claws, and keen teeth, then they would conceal nothing from me. Now the lion was very old, ferocious and large, his yellow hair hung over his neck, he appeared quite unconquerable, so that I was almost afraid of my temerity and would gladly have turned back if my promise and also the circumstance that the elders stood about me and were waiting to see what I would do, had allowed me to give up. In great confidence I approached the lion in his den and began to caress him, but he looked at me so fiercely with his brightly shining eyes that I could hardly restrain my tears. Just then I remembered that I had learned from one of the elders, while we were going to the lion's den, that very many people had undertaken to overcome the lion and very few could accomplish it. I was unwilling to be disgraced, and I recalled several grips that I had learned with great diligence in athletics, besides which I was well versed in natural magic [magia] so I gave up the caresses and seized the lion so dextrously, artfully and subtlely, that before he was well aware of it I forced the blood out of his body,

yea, even out of his heart. It was beautifully red but very choleric. I dissected him further and found, a fact which caused me much wonder, that his bones were white as snow and there was much more bone than there was blood.

[7]. Now when my dear elders, who stood above around the den and looked at me, were aware of it, they disputed earnestly with each other, for so much I could infer from their motions but what they said I could not hear since I was deep down in the den. Yet as they came close in dispute I heard that one said, " He must bring him to life again, else he can not be our colleague." I was unwilling to undertake further difficulties, and betook myself out of the den to a great place, and came, I know not how, on a very high wall, whose height rose over 100 ells towards the clouds, but on top was not one foot wide. And there went up from the beginning, where I ascended, to the end an iron hand rail right along the center of the wall, with many leaded supports. On this wall I came, I say, and meseems there went on the right side of the railing a man several paces before me.

[8]. But as I followed him awhile I saw another following me on the other side, yet it was doubtful whether man or woman, that called to me and said that it was better walking on his side than where I went, as I readily believed, because the railing that stood near the middle made the path so narrow that the going at such a height was very bad. Then I saw also some that wished to go on that path, fall down below behind me, therefore I swung under the railing holding tight with my hands and went forward on the other [left] side, till I finally came to a place on the wall which was very precipitous and dangerous to descend. Then first I repented that I had not stayed on the other [right] side and I could not go under to the other side as it was also impossible to turn round and get on the other path. So I

risked it, trusted to my good feet, held myself tight and came down without harm, and as I walked a little further, looked and knew of no other danger, but also knew not what had become of wall and railing.

[9]. After I came down, there stood in that place a beautiful rose bush, on which beautiful red and white roses were growing, the red more numerous, however, than the white. I broke off some roses from the bush and put them on my hat. But there seemed to be in the same place a wall, surrounding a great garden. In the garden were lads, and their lasses who would gladly be in the garden, but would not wander widely, or take the trouble to come to the gates. So I pitied them. I went further along the path by which I had come, still on the level, and went so fast that I soon came to some houses, where I supposed I should find the gardener's house. But I found there many people, each having his own room. They were slow. Two together they worked diligently, yet each had his own work. [The meaning may be either that working alone they were slow, but in twos they worked diligently; or two of them worked together and were diligent. Both amount to the same thing as we shall later realize.] But what they did, it seems, I had myself done before and all their work was familiar to me. Especially, thought I, see, if so many other people do so much dirty and sloppy work, that is only an appearance according to each one's conceit, but has no reason in Nature, so it may also be pardoned in you. I wished, therefore, because I knew such tricks vanished like smoke, to remain here no longer in vain and proceeded on my former way.

[10]. After I had arrived at the gate of the garden, some on one side looked sourly at me, so that I was afraid they might hinder me in my project; but others said, " See, he will into the garden, and we have done garden service here

so long, and have never gotten in; we will laugh him down if he fails." But I did not regard all that, as I knew the conditions of this garden better than they, even if I had never been in it, but went right to a gate that was tight shut so that one could neither see nor find a keyhole. I noticed, however, that a little round hole that with ordinary eyes could not be seen, was in the door, and thought immediately, that must be the way the door is opened, was ready with my specially prepared Diederich, unlocked and went in. When I was inside the door, I found several other bolted doors, which I yet opened without trouble. Here, however, was a passage way, just as if it was in a well built house, some six feet wide and twenty long, with a roof above. And though the other doors were still locked, I could easily see through them into the garden as the first door was open.

[11]. I wandered into the garden in God's name, and found in the midst of it a small garden, that was square and six roods long, hedged in with rose thorns, and the roses bloomed beautifully. But as it was raining gently, and the sun shone in it, it caused a very lovely rainbow. When I had passed beyond the little garden and would go to the place where I was to help the maids, behold I was aware that instead of the walls a low hurdle stood there, and there went along by the rose garden the most beautiful maiden arrayed in white satin, with the most stately youth, who was in scarlet each giving arm to the other, and carrying in their hands many fragrant roses. I spoke to them and asked them how they had come over the hurdle. "This, my beloved bridegroom," said she, "has helped me over, and we are going now out of this beautiful garden into our apartment to enjoy the pleasures of love." "I am glad," said I, "that without any further trouble on my part your desires are satisfied; yet see how I have hurried, and have run so long a way in so short a time to serve you." After that

I came into a great mill built inside of stones, in which were no flour bins or other things that pertained to grinding but one saw through the walls several water wheels going in water. I asked why it had equipment for grinding. An old miller answered that the mill was shut down on the other side. Just then I also saw a miller's boy go in from the sluice plank [Schutzensteg], and I followed after him. When I had come over the plank [Steg], which had the water wheels on the left, I stood still and was amazed at what I saw there. For the wheels were now higher than the plank, the water coal black, but its drops were yet white, and the sluice planks were not over three fingers wide. Still I ventured back and held onto the sticks that were over the sluice planks and so came safely and dry over the water. Then I asked the old miller how many water wheels he had. " Ten," answered he. The adventure stuck in my mind. I should have gladly known what the meaning was. But as I noticed that the miller would not leave I went away, and there was in front of the mill a lofty paved hill, on which were some of the previously mentioned elders who walked in the sun, which then shone very warm, and they had a letter from the whole faculty written to them, on which they were consulting. [In our modern mode of expression, the elders had directed a letter to the sun, and so I find the passage in an English version of the parable. This generally bungling translation is nevertheless not in the least authoritative. And although an acceptable meaning is derived from it, if one regards the sun as the just mentioned " prince," yet I believe a freer translation should be given . . . the elders walked in the warm sunshine; they consulted about a letter written to them by the faculty.] I soon noticed what the contents must be, and that it concerned me. I went therefore to them and said, " Gentlemen, does it concern me?" " Yes," said they, " you must keep in marriage the woman that you have recently taken or we must notify our prince." I said, " that is no trouble as I was born at the same time as she and brought up as a child with her, and as I have taken her once I will keep her forever, and death itself shall not part us, for I have an ardent affection for her." "What have we then to complain of?" replied they. "The bride is content, and we have your will; you must copulate." "Contented," said I. "Well," said one, "the lion will then regain his life and become more powerful and mighty than before."

[12]. Then occurred to me my previous trouble and labor and I thought to myself that for particular reasons it must not concern me but some other that is well known to me; then I saw our bridegroom and his bride go by in their previous attire, ready and prepared for copulation, which gave me great joy, for I was in great distress lest the thing might concern me.

[13]. When, then, as mentioned, our bridegroom in his brilliant scarlet clothes with his dearest bride, whose white satin coat shot forth bright rays, came to the proper marriage age, they joined the two so quickly that I wondered not a little that this maid, that was supposed to be the mother of the bridegroom, was still so young that she appeared to be just born.

[14]. Now I do not know what sin these two must have committed except that although they were brother and sister, they were in such wise bound by ties of love, that they could not be separated, and so, as it were, wished to be punished for incest. These two were, instead of a bride bed and magnificent marriage, condemned and shut up in an enduring and everlasting prison, which, because of their high birth and goodly state, and also so that in future they should not be guilty in secret, but all their conduct should be known to the guard placed over them and in his sight, was made

quite transparent, bright and clear like a crystal, and round
like a sphere of heaven, and there they were with continual
tears and true contrition to atone and make reparation for
their past misdeeds. [Instead of to a bride bed the two
were brought to a prison, so that their actions could be
watched. The prison was transparent; it was a bright
crystal clear chamber, like a sphere of heaven, corresponding
to the high position of the two persons.] Previously, however, all their rare clothing and finery that they had put on
for ornament was taken away, so that in such a chamber
they must be quite naked and merely dwell with each other.
[It is not directly understood by these words that a cohabitation in modern sense (coition) is meant. According to modern language the passage must be rendered, " had to dwell
near each other naked and bare." One is reminded, moreover, of the nuptial customs that are observed particularly
in the marriage of persons of high birth. In any case and,
in spite of my reservation, what occurs is conducive or designed to lead to the sexual union.] Besides they gave them
no one that had to go into the chamber to wait on them,
but after they put in all the necessities in the way of meat
and drink, which were created from the afore mentioned
water, the door of the chamber was fast bolted and locked,
the faculty seal impressed on it and I was enjoined that I
should guard them here, and spend the winter before the
door; the chamber should be duly warmed so that they be
neither too hot nor too cold, and they could neither come
out nor escape. But should they, on account of any hope
of breaking this mandate, escape, I would thereupon be
justly subjected to heavy punishment. I was not pleased
by the thing, my fear and solicitude made me faint hearted,
for I communed with myself that it was no small thing that
had befallen me, as I knew also that the college of wisdom
was accustomed not to lie but to put into action what it

said. Yet because I could not change it, beside which this locked chamber stood in the center of a strong tower and surrounded with strong bulwarks and high walls, in which one could with a small but continuous fire warm the whole chamber, I undertook this office, and began in God's name to warm the chamber, and protect the imprisoned pair from the cold. But what happened? As soon as they perceived the slightest warmth they embraced each other so tenderly that the like will not soon be seen, and stayed so hot that the young bridegroom's heart in his body dissolved for ardent love, also his whole body almost melted in his beloved's arms and fell apart. When she who loved him no less than he did her, saw this, she wept over him passionately and, as it were buried him with her tears so that one could not see, for her gushing tears that overflowed everything, where he went. Her weeping and sorrowing had driven her to this in a short time, and she would not for deep anguish of heart live longer, but voluntarily gave herself to death. Ah woe is me. In what pain and need and trouble was I that my two charges had quite disappeared in water, and death alone was left for me. My certain destruction stood before my eyes, and what was the greatest hardship to me, I feared the threatened shame and disgrace that would happen to me, more than the injury that would overtake me.

[15]. As I now passed several days in such solitude and pondered over the question how I could remedy my affairs, it occurred to me how Medea had revived the dead body of Aeson, and I thought to myself, "If Medea could do such thing, why should such a thing fail me?" I began at once to bethink me how I would do it, found however no better way than that I should persist with continual warmth until the waters disappeared, and I might see again the corpses of our lovers. As I hoped to come off without danger and with great advantage and praise I went on with my warmth

that I had begun and continued it forty whole days, as I was aware that the water kept on diminishing the longer I kept it up, and the corpses that were yet as black as coal, began again to be visible. And truly this would have occurred before if the chamber had not been all too securely locked and bolted. Which I yet did not avail to open. For I noted particularly that the water that rose and hastened to the clouds, collected above in the chamber and fell down like rain, so that nothing could come of it, until our bridegroom with his dearest bride, dead and rotten, and therefore hideously stinking, lay before my eyes.

All the while the sunshine in the moist weather caused an exceedingly beautiful rainbow to be seen, in the chamber, with surprisingly beautiful colors, which overjoyed not a little my overpowering affliction. Much more was I delighted that I saw my two lovers lying before me again. But as no joy is so great but is mixed with much sadness, so I was troubled in my joy thinking that my charges lay still dead before me, and one could trace no life in them. But because I knew that their chamber was made of such pure and thick material, also so tight-locked that their soul and spirit could not get out, but was still closely guarded within, I continued with my steady warmth day and night, to perform my delegated office, quite impressed with the fact that the two would not return to their bodies, as long as the moisture continued. For in the moist state nature keeps itself the same, as I then also found in fact and in truth. For I was aware upon careful examination that from the earth at evening through the power of the sun, many vapors arose and drew themselves up just as the sun draws water. They were condensed in the night in a lovely and very fruitful dew, which very early in the morning fell and moistened the earth and washed our dead corpses, so that from day to day, the longer such bathing and washing continued, the more beautiful and whiter they became. But the fairer and whiter they became, the more they lost moisture, till finally the air being bright and beautiful, and all the mist and moist weather, having passed, the spirit and soul of the bride could hold itself no longer in the bright air, but went back into the clarified and still more transfigured body of the queen, who soon experienced it [i.e. her soul and spirit] and at once lived again. This, then, as I could easily observe, not a little pleased me, especially as I saw her arise in surpassingly costly garments whose like was never seen on earth, and with a precious crown decked with bright diamonds; and also heard her speak. " Hear ye children of men and perceive ye that are born of women, that the most high power can set up kings and can remove kings. He makes rich and poor, according to his will. He kills and makes again to live."

[16]. See in me a true and living example of all that. I was great and became small, but now after having been humbled, I am a queen elevated over many kingdoms. I have been killed and made to live. To poor me have been trusted and given over the great treasures of the sages and the mighty.

[17]. "Therefore power is also given me to make the poor rich, show kindness to the lowly, and bring health to the sick. But I am not yet like my well-beloved brother, the great and powerful king, who is still to be awakened from the dead. When he comes he will prove that my words are true."

[18]. And when she said that the sun shone very bright, and the day was warmer than before, and the dog days were at hand. But because, a long time before, there were prepared for the lordly and great wedding of our new queen many costly robes, as of black velvet, ashen damask, gray silk, silver taffeta, snow white satin, even one studded with

surpassingly beautiful silver pieces and with precious pearls and lordly bright-gleaming diamonds, so likewise different garments were arranged and prepared for the young king, namely of carnation, yellow Auranian colors, precious gear, and finally a red velvet garment with precious rubies and thickly incrusted with carbuncles. But the tailors that made their clothes were quite invisible, so that I also wondered as I saw one coat prepared after another and one garment after another, how these things came to pass, since I well knew that no one came into the chamber except the bride-groom with his bride. So that what I wondered at most of all was that as soon as another coat or garment was ready, the first immediately vanished before my eyes, so that I knew not whence they came or who had taken them away.

[19]. When now this precious clothing was ready, the great and mighty king appeared in great splendor and magnif-icence, to which nothing might be compared. And when he found himself shut in, he begged me with friendly and very gracious words, to open the door for him and permit him to go out; it would prove of great advantage to me. Although I was strictly forbidden to open the chamber, yet the grand appearance and the winning persuasiveness of the king disconcerted me so that I cheerfully let him go. And when he went out he was so friendly and so gracious and yet so meek that he proved indeed that nothing did so grace high persons as did these virtues.

[20]. But because he had passed the dog days in great heat, he was very thirsty, also faint and tired and directed me to dip up some of the swift running water under the mill wheels, and bring it, and when I did, he drank a large part with great eagerness, went back into his chamber, and bade me close the door fast behind him, so that no one might disturb him or wake him from sleep.

[21]. Here he rested for a few days and called to me to

open the door. Methought, however, that he was much more beautiful, more ruddy and lordly, which he then also remarked and deemed it a lordly and wholesome water, drank much of it, more than before so that I was resolved to build the chamber much larger. [Evidently because the inmate increased in size.] When now the king had drunk to his satisfaction of this precious drink, which yet the unknowing regard as nothing he became so beautiful and lordly, that in my whole life I never saw a more lordly person nor more lordly demeanor. Then he led me into his kingdom, and showed me all the treasures and riches of the world, so that I must confess, that not only had the queen announced the truth, but also had omitted to describe the greater part of it as it seemed to those that know it. For there was no end of gold and noble carbuncle there; rejuvenation and restoration of natural forces, and also recovery of lost health, and removal of all diseases were a common thing in that place. The most precious of all was that the people of that land knew their creator, feared and honored him, and asked of him wisdom and understanding, and finally after this transitory glory an everlasting blessedness. To that end help us God the Father, Son and Holy Ghost. Amen.

The author of the preceding narrative calls it a parable. Its significance may have indeed appeared quite transparent to him, and he presupposes that the readers of his day knew what form of learning he masked in it. The story impresses us as rather a fairy story or a picturesque dream. If we compare parables that come nearer to our modern point of view and are easily understood on account of their simplicity, like those of Ruckert or those of the New Testament, the difference can be clearly seen. The unnamed author evidently pursues a definite aim; one does find some unity in the bizarre confusion of his ideas; but what he is aiming at and what he wishes to tell us with his images we cannot immediately conceive. The main fact for us is that the anonymous writer speaks in a language that shows decided affinity with that of dreams and myths. Therefore, however we may explain in what follows the peculiarly visionary character of the parable, we feel compelled to examine it with the help of a psychological method, which, endeavoring to get from the surface to the depths, will be able to trace analytically the formative powers of the dream life and allied phenomena, and explain their mysterious symbols.

I have still to reveal in what book and in what circumstances the parable appears. It is in the second volume of a book " Geheime Figuren der Rosenkreuzer aus dem 16ten und 17ten Jahrhundert," published at Altona about 1785–90. Its chief contents are large plates with pictorial representations and with them a number of pages of text. According to a note on the title page, the contents are " for the first time brought to light from an old manuscript." The parable is in the second volume of a three-volume series which bears the subtitle: Ein güldener Tractat vom philosophischen Steine. Von einem noch lebenden, doch ungenannten Philosopho, den Filiis doctrinae zur Lehre, den Fratribus Aureae Crucis aber zur Nachrichtung beschrieben. Anno, M.D.C.XXV.

If I add that this book is an hermetic treatise (alchemistic), it may furnish a general classification for it, but will hardly give any definite idea of its nature, not merely on account of the oblivion into which this kind of writing has now fallen, but also because the few ideas usually connected with it produce a distorted picture.   The hermetic art, as it is treated here, the principles of which strike us to-day as fantastic, is related to several " secret " sciences and organizations, some of which have been discredited: magic, kabbala, rosicrucianism, etc.   It is particularly closely connected with alchemy so that the terms " hermetic art " and " alchemy " (and even " royal art ") are often used synonymously. This " art "— to call it by the name that not without some justification it applies to itself — leads us by virtue of its many ramifications into a large number of provinces, which furnish us desirable material for our research.

So I will first, purposely advancing on one line, regard the parable as a dream or a fairy tale and analyze it psychoanalytically.   This treatment will, for the information of the reader, be preceded by a short exposition of psychoanalysis as a method of interpretation of dreams and fairy tales.   Then I will, still seeking for the roots of the matter, introduce the doctrines that the pictorial language of the parable symbolizes.   I will give consideration to the chemical viewpoint of alchemy and also the hermetic philosophy and its hieroglyphic educational methods.

Connections will be developed with religious and

ethical topics, and we shall have to take into account the historical and psychological relations of hermetic thought with rosicrucianism in its various forms, and freemasonry. And when we begin, at the conclusion of the analytical section of my work, to apply to the solution of our parable and of several folk tales the insight we have gained, we shall be confronted with a problem in which we shall face two apparently contradictory interpretations, according to whether we follow the lead of psychoanalysis or of the hermetic, hieroglyphic solution. The question will then arise whether and how the contradiction occurs. How shall we bring into relation with each other and reconcile the two different interpretations which are quite different and complete in themselves?

The question arising from the several illustrations expands into a general problem, to which the synthetic part of my book is devoted. This will, among other considerations, lead us into the psychology of symbol-making where again the discoveries of psychoanalysis come to our aid. We shall not be satisfied with analysis, but endeavor to follow up certain evolutionary tendencies which, expressed in psychological symbols, developing according to natural laws, will allow us to conjecture a spiritual building up or progression that one might call an anabasis. We shall see plainly by this method of study how the original contradiction arises and how what was previously irreconcilable, turns out to be two poles of an evolutionary process. By that means, several principles of myth interpretation will be derived.

I have just spoken of an anabasis. By that we are to understand a forward movement in a moral or religious sense. The most intensive exemplar of the anabasis (whatever this may be) is mysticism. I can but grope about in the psychology of mysticism; I trust I may have more confidence at that point where I look at its symbolism from the ethical point of view.

## SECTION II

# DREAM AND MYTH INTERPRETATION

[Readers versed in Freud's psychoanalysis are requested to pass over this chapter as they will find only familiar matter.]

IN the narrative which we have just examined its dream-like character is quite noticeable. On what does it depend? Evidently the Parable must bear marks that are peculiar to the dream. In looking for correspondences we discover them even upon superficial examination.

Most noticeable is the complete and sudden change of place. The wanderer, as I will hereafter call the narrator of the parable, sees himself immediately transported from the place near the lion's den to the top of a wall, and does not know how he has come there. Later he comes down just as suddenly. And in still other parts of the story there occurs just as rapid changes of scene as one is accustomed to in dreams. Characteristic also is the fact that objects change or vanish; the shift of scene resembles also, as often in a dream, a complete transformation. Thus, for instance, as soon as the wanderer has left the wall, it vanishes without leaving a trace, as if it had never been. A similar change is also required in the garden scene where, instead of the previously observed enclosing-wall, a low hedge appears in a surprising manner.

Further, we are surprised by instances of knowledge without perception. Often in a dream one knows something without having experienced it in person. We simply know, without knowing how, that in such a house something definite and full of mystery has happened; or we know that this man, whom we see now for the first time, is called so and so; we are in some place for the first time but know quite surely that there must be a fountain behind that wall to which for any reason we have to go, etc. Such unmediated knowledge occurs several times in the parable. In the beginning of the narrative the wanderer, although a stranger, knows that the lovely meadow is called by its inhabitants Pratum felicitatis. He knows intuitively the name of one of the men unknown to him. In the garden scene he knows, although he has noticed only the young men, that some young women (whom on account of the nature of the place he cannot then see) are desirous of going into the garden to these young men. One might say that all this is merely a peculiarity of the representation inasmuch as the author has for convenience, or on account of lack of skill, or for brevity, left out some connecting link which would have afforded us the means of acquiring this unexplained knowledge. The likeness to the dream therefore would in that case be inadmissible. To this objection it may be replied, that the dream does exactly like the author of the parable. Our study is chiefly con-

cerned with the product of the fancy and forces us
to the observation (whatever may be the cause of it)
that the parable and the dream life have certain " peculiarities of representation " in common.

In contrast to the miraculous knowledge we find
in the dream a peculiar unsureness in many things,
particularly in those which concern the personality of
the wanderer.  When the elders inform the wanderer that he must marry the woman that he has
taken, he does not know clearly whether the matter at all concerns him or not; a remarkable fluctu-
ation in his attitude takes place.  We wonder
whether he has taken on the rôle of the bridegroom
or, quite the reverse, the bridegroom has taken the
wanderer's.  We are likewise struck by similar un-
certainties, like those during the walk on top of the
wall where the wanderer is followed by some one,
of whom he does not know whether it is a man or a
woman.  Here belong also those passages of the
narrative introduced by the wanderer with " as if,"
etc.  In the search for the gardener's house he
chances upon many people and " it seems " that he
has himself done what these people are there doing.

Quite characteristic also are the different obstructions and other difficulties placed in the path of the
wanderer.  Even in the first paragraph of the narrative we hear that he is startled, would gladly turn
back, but cannot because a strong wind prevents him.
On top of the wall the railing makes his progress
difficult; on other occasions a wall, or a door.  The
first experience, especially, recalls those frequent oc-

currences in dreams where, anxiously turning in
flight or oppressed by tormenting haste, we cannot
move.   In connection with what is distressing and
threatening, as described in the precipitous slope of
the wall and the narrow plank by the mill, belong
also the desperate tasks and demands — quite usual
in dreams and myths — that meet the wanderer.
Among such tasks or dangers I will only mention
the severe examination by the elders, the struggle
with the lion, the obligation to marry, and the burden
of responsibility for the nuptial pair, all of which
cause the wanderer so much anxiety.

Among the evident dream analogies belongs finally
(without, however, completing my list of them) the
peculiar logic that appears quite conventional to the
wanderer or the dreamer, but seldom satisfies the
reader or the careful reasoner.   As examples, I
mention that the dead lion will be called to life
again if the wanderer marries the woman that he
recently took; and that they put the two lovers that
they want to punish for incest, after they have carefully removed all the clothes from their bodies, into
a prison where these lovingly embrace.

So much for the external resemblances of the
parable with the dream life.   The deeper affinity
which can be shown in its innermost structure will
first appear in the psychoanalytic treatment.   And
now it will be advisable for me to give readers not
intimately acquainted with dream psychology some
information concerning modern investigations in
dream life and in particular concerning psychoana-

lytic doctrines and discoveries.   Naturally I can do
this only in the briefest manner.   For a more thorough study I must refer the reader to the work of
Fɪeud and his school.   The most important books
are mentioned in the bibliography at the end.

Modern scientific investigation of dreams, in
which Freud has been a pioneer, has come to the
conclusion, but in a different sense from the popular
belief, that dreams have a significance.   While the
popular belief says that they foretell something of
the future, science shows that they have a meaning
that is present in the psyche and determined by the
past.   Dreams are then, as Freud's results show,
always wish phantasies.   [I give here only exposition, not criticism.   My later application of psychoanalysis will show what reservations I make concerning Freud's doctrines.]   In them wishes, strivings,
impulses work themselves out, rising to the surface
from the depths of the soul.   When they come in
waking life, wish phantasies are sometimes called
castles in the air.   In dreams we have the fulfillment
of wishes that are not or cannot be fulfilled.

But the impulses that the dreams call up are principally such wishes and impulses as we cannot ourselves acknowledge and such as in a waking state we
reject as soon as they attempt to announce themselves, as for instance, animal tendencies or such
sexual desires as we are unwilling to admit, and also
suppressed or " repressed " impulses.   As a result
of being repressed they have the peculiarity of being
in general inaccessible to consciousness.   [Freud

speaks particularly of crassly egoistic actuations. The criminal element in them is emphasized by Stekel.]

One not initiated into dream analysis may object that the obvious evidence is against this theory. For the majority of dreams picture quite inoffensive processes that have nothing to do with impulses and passions which are worthy of rejection on either moral or other grounds.  The objection appears at first sight to be well founded, but collapses as soon as we learn that the critical power of morality, which does not desert us by day, retains by night a part of its power; and that therefore the fugitive impulses and tendencies that seek the darkness and dare not come forth by day, dare not even at night unveil their true aspect but have to approach, as it were, in costumes, or disguised as symbols or allegories, in order to pass unchallenged.  The superintending power, that I just now called the power of morality, is compared very pertinently to a censor.  What our psyche produces is, so to speak, subjected to a censor before it is allowed to emerge into the light of consciousness.  And if the fugitive elements want to venture forth they must be correspondingly disguised, in order to pass the censor.  Freud calls this disguising or paraphrasing process the dream disfigurement.  The literal is thereby displaced by the figurative, an allusion intimated through a nebulous atmosphere.  Thus, in the following example, an unconscious death wish is exhibited.  In the examination of a lady's dream it struck me that the motive of a dead child occurred repeatedly, generally in connection with picnics. During an analysis the lady observed that when she was a girl the children, her younger brothers and sisters, were often the obstacles when it was proposed to have a party or celebration or the like. The association Kinder (= children) Hinderniss (= obstacle) furnished the key to a solution of the stereotyped dream motive. As further indications showed, it concerned the children of a married man whom she loved. The children prevented the man separating from his wife in order to marry the lady. In waking life she would not, of course, admit a wish for the death of the embarrassing children, but in dreams the wish broke through and represented the secretly wished situation. The children are dead and nothing now stands in the way of the " party " or the celebration (wedding). The double sense of the word " party " is noticeable. (In German " eine Partie machen " means both to go on an excursion and to make a matrimonial match.) Such puns are readily made use of by dreams, in order to make the objectionable appear unobjectionable and so to get by the censor.

Psychoanalytic procedure, employed in the interpretation of dreams of any person can be called a scientifically organized confession that traces out with infinite patience even to the smallest ramifications, the spiritual inventory of what was tucked away in the mind of the person undergoing it. Psychoanalysis is used in medical practice to discover

and relieve the spiritual causes of neurotic phenomena.   The patient is induced to tell more and
more, starting from a given point, thereby going
into the most intimate details, and yet we are aware,
in the network of outcropping thoughts and memories, of certain points of connection, which have
dominating significance for the affective life of the
person being studied.   Here the path begins to be
hard because it leads into the intimately personal.
The secret places of the soul set up a powerful opposition to the intruder, even without the purposive
action of the patient.   Right there are, however, so
to speak, the sore spots (pathogenic " complexes ")
of the psyche, towards which the research is directed.
Firmly advancing in spite of the limitations, we lay
bare these roots of the soul that strive to cling to
the unconscious.   Those are the disfigured elements
just mentioned; all of the items of the spiritual
inventory from which the person in question has toil-
somely " worked himself out " and from which he
supposes himself free.   They must be silent because
they stand in some contradictory relation to the character in which the person has clothed himself; and
if they, the subterranean elements still try to announce themselves, he hurls them back immediately
into their underworld; he allows himself to think of
nothing that offends too much his attitudes, his
morality and his feelings.   He does not give verbal
expression to the disturbers of the peace that dwell
in his heart of hearts.

The mischief makers are, however, merely repressed, not dead. They are like the Titans [On this similarity rests the psychologic term "titanic," used frequently by me in what follows.] which were not crushed by the gods of Olympus, but only shut up in the depths of Tartarus. There they wait for the time when they can again arise and show their faces in daylight. The earth trembles at their attempts to free themselves. Thus the titanic forces of the soul strive powerfully upward. And as they may not live in the light of consciousness they rave in darkness. They take the main part in the procreation of dreams, produce in some cases hysterical symptoms, compulsion ideas and acts, anxiety neuroses, etc. The examination of these psychic disturbances is not without importance for our later researches.

Psychoanalysis, which has not at any time been limited to medical practice, but soon began with its torch to illumine the activity of the human spirit in all its forms (poetry, myth-making, etc.), was decried as pernicious in many quarters. [The question as to how widely psychoanalysis may be employed would at this time lead us too far, yet it will be considered in Sect. 1, of the synthetic part of this volume.] Now it is indeed true that it leads us toward all kinds of spiritual refuse. It does so, however, in the service of truth, and it would be unfortunate to deny to truth its right to justify itself. Any one determined to do so could in that case defend a theory that sexual maladies are acquired by catching a cold.

The spiritual refuse that psychoanalysis uncovers is like the manure on which our cultivated fruits thrive. The dark titanic impulses are the raw material from which in every man, the work of civilization forms an ethical character. Where there is a strong light there are deep shadows. Should we be so insincere as to deny, because of supposed danger, the shadows in our inmost selves? Do we not diminish the light by so doing? Morality, in whose name we are so scrupulous, demands above everything else, truth and sincerity. But the beginning of all truth is that we do not impose upon ourselves. " Know thyself " is written over the entrance of the Pythian sanctuary. And it is this inspiring summons of the radiant god of Delphi that psychoanalysis seeks to meet.

After this introductory notice, it will be possible properly to understand the following instructive example, which contains exquisite sexual symbolism.

Dream of Mr. T. " I dreamed I was riding on the railroad. Near me sat a delicate, effeminate young man or boy; his presence caused erotic feelings in me to a certain extent. (It appeared as if I put my arm about him.) The train came to a standstill; we had arrived at a station and got out. I went with the boy into a valley through which ran a small brook, on whose bank were strawberries. We picked a great many. After I had gathered a large number I returned to the railway and awoke."

Supplementary communication. "I think I re-member that an uncomfortable feeling came over me in the boy's company. The valley branched off to the left from the railway."

From a discussion of the dream it next appeared that T., who, as far as I knew, entertained a pro-nounced aversion to homosexuality, had read a short time before a detailed account of a notorious trial then going on in Germany, that was concerned with real homosexual actions. [In consciousness, of course. In the suppressed depths of unconsciousness the infantile homosexual component also will surely be found.]   An incident from it, probably supported by some unconscious impulse, crowded its way into the dream as an erotic wish, hence the affectionate scene in the railway train.   So far the matter would be intelligible even if in an erotic day dream the image of a boy, considering the existing sexual tend-ency of T., had been resolutely rejected by him. How are the other processes in the dream related to it?   Do they not at first sight appear unconnected or meaningless?

And yet in them are manifested the fulfillment of the wish implied in the erotic excitement in the com-pany of the boy. The homosexual action of this wish fulfillment would have been insufferable to the dream censor; it must be intimated symbolically. And the remainder of the dream is accordingly noth-ing but a dextrous veiling of a procedure hostile to the censor.

Even that the train comes to a standstill is a polite paraphrase. [Paraphrase as the dreamer communicated to me, of an actual physical condition — an erection.] Similar meaning is conveyed by the word station, which reminds us of the Latin word status (from stare, to stand). The scene in the car recalls moreover the joke in a story which often used to occur to T. "A lady invited to a reception, where there were also young girls, a *Hungarian* [accentuated now, on account of what follows] (the typical Vienna joker), who is feared on account of his racy wit. She enjoined him at the same time, in view of the presence of the girls, not to treat them to any of his spicy jests. The Hungarian agreed and appeared at the party. To the amazement of the lady, he proposed the following riddle: 'One can enter from in front, or from behind, only one has to stand up.' Observing the despair of the lady, he, with a sly, innocent look, said, 'But well then, what is it? Simply a trolley car.' Next day the daughter of the house appeared before her schoolmates in the high school with the following: 'Girls, I heard a great joke yesterday; one can go in from in front or behind, only one must be stiff.'" [A neat contribution, by the way, to the psychology of innocent girlhood.] The anecdote was related to T. by a man later known to him as a homosexual. T. had been with few Hungarians, but with these few, homosexuality had been, as it happened, a favorite subject of conversation.

In the above we find many highly suggestive elements. The most suggestive is, however, the straw-berries. T. had, as appeared during the process of the analysis, a couple of days before the dream read a French story where the expression (new to him) *cueillir des fraises* occurred. He went to a Frenchman for the explanation of this phrase and learned that it was a delicate way of speaking of the sexual act, because lovers like to go into the woods under the pretext of picking strawberries, and thus separate themselves from the rest of the company. In whatever way the dream wish conceived its gratification, the valley (between the two hills!) through which the brook flowed furnishes a quite definite suggestion. Here also the above mentioned " from behind " probably gets a meaning.

The circumstance that the dream has, as it were, two faces, with one that it openly exposes to view, implies that a distinction must be made between the manifest and the latent material. The openly exposed face is the manifest dream content (as the wording of the dream report represents the dream) ; what is concealed is the latent dream thoughts. For the most part a broad tissue of dream thoughts is condensed into a dream. A part of the dream thoughts (not all) belongs regularly to the titanic elements of our psyche. The shaping of the dream out of the dream thoughts is called by Freud the dream work. Four principles direct it, Condensation, Displacement, Representability, and Secondary Elaboration.

Condensation was just now mentioned. Many dream thoughts are condensed to relatively few, but therefore all the more significant, images. Every image (person, object, etc.) is wont to be " determined " by several dream thoughts. Hence we speak of multiple determination or " Overdetermination."

Displacement shows itself in the fact that the dream (evidently in the service of distortion) pushes forward the unreal and pushes aside the real; in short, rearranges the psychic values (interest) in such a way that the dream in comparison with its latent thoughts appears as it were displaced or " elsewhere centered."

As the dream is a perceptual representation it must put into perceptually comprehensible form everything that it wants to express, even that which is most abstract. The tendency to vividly perceptual or plastic expression that is characteristic of the dream, corresponds accordingly to the Representability.

To the Secondary Elaboration we have to credit the last polishing of the dream fabric. It looks after the logical connection in the pictorial material, which is created by the displacing dream work. " This function (i.e., the secondary elaboration) proceeds after the manner which the poet maliciously ascribes to the philosopher; with its shreds and patches it stops the gaps in the structure of the dream. The result of its effort is that the dream loses its appearance of absurdity and disconnectedness and approaches the standard of a comprehensible experience. But the effort is not always crowned with complete success." (Freud, "Traumdeutung," p. 330.) The secondary elaboration can be compared also to the erection of a façade.

Of the entire dreamwork Freud says ("Traumdeutung," p. 338) comprehensively that it is "not merely more careless, more incorrect, more easily forgotten or more fragmentary than waking thought; it is something qualitatively quite different and therefore not in the least comparable with it. It does not, in fact, think, reckon, or judge, but limits itself to remodeling. It may be exhaustively described if we keep in view the conditions which its productions have to satisfy. These productions, the dream, will have first of all to avoid the censor, and for this purpose the dream work resorts to displacement of psychic intensities even to the point of changing all psychic values; thoughts must be exclusively or predominantly given in the material of visual and auditory memory images, and from this grows that demand for representability which it answers with new displacements. Greater intensities must apparently be attained here, than are at its disposal in dream thoughts at night, and this purpose is served by the extreme condensation which affects the elements of the dream thoughts. There is little regard for the logical relations of the thought material; they find finally an indirect representation in formal peculiarities of dreams. The affects of dream thoughts suffer slighter changes than their image

content. They are usually repressed. Where they
are retained they are detached from images and
grouped according to their similarity."

Briefly to express the nature of the dream, Stekel
gives in one place (" Sprache des Traumes," p. 107)
this concise characterization: " The dream is a
play of images in the service of the affects."

A nearly exact formula for the dream has been
contributed by Freud and Rank, " On the foundation and with the help of repressed infantile sexual
material, the dream regularly represents as fulfilled
actual wishes and usually also erotic wishes in disguised and symbolically veiled form." (Jb.; ps. F.,
p. 519, and Trdtg., p. 117.) In this formula the
wish fulfillment, following Freud's view, is prepon-
derant, yet it would appear to me that it is given too
exclusive a rôle in the (chiefly affective) background
of the dream. An important point is the infantile in
the dream, in which connection we must mention
the Regression.

Regression is a kind of psychic retrogression that
takes place in manifold ways in the dream (and
related psychic events). The dream reaches back
towards infantile memories and wishes. [Some-
times this is already recognizable in the manifest
dream content. Usually, however, it is first disclosed by psychoanalysis. Strongly repressed, and
therefore difficult of access, is this infantile sexual
material. On the infantile forms of sexuality, see
Freud, " Three Contributions to Sexual Theory."]
It reaches back also from the complicated and com-

pleted to a more primitive function, from abstract
thought to perceptual images, from practical activity to hallucinatory wish fulfillment.    [The latter
with especial significance in the convenience dreams.
We fall asleep, for instance, when thirsty, then instead of reaching for the glass of water, we dream
of the drink.]    The dreamer thus approaches his
own childhood, as he does likewise the childhood of
the human race, by reaching back for the more primitive perceptual mode of thought.    [On the second
kind of regression the Zurich psychiatrist, C. G.
Jung, has made extraordinary interesting revelations.    His writings will further occupy our attention later.]

Nietzsche writes (" Menschliches, Allzumensch-
liches "), " In sleep and in dreams we pass through
the entire curriculum of primitive mankind. . . . I
mean as even to-day we think in dreams, mankind
thought in waking life through many thousand years;
the first cause that struck his spirit in order to explain anything that needed explanation satisfied him
and passed as truth.    In dreams this piece of ancient humanity works on in us, for it is the germ
from which the higher reason developed and in
every man still develops.    The dream takes us back
into remote conditions of human culture and puts in
our hand the means of understanding it better.    The
dream thought is now so easy because, during the
enormous duration of the evolution of mankind we
have been so well trained in just this form of cheap,
phantastic explanation by the first agreeable fancy.

In that respect the dream is a means of recovery for the brain, which by day has to satisfy the strenuous demands of thought required by the higher culture." (Works, Vol. II, pp. 27 ff.)

If we remember that the explanation of nature and the philosophizing of unschooled humanity is consummated in the form of myths, we can deduce from the preceding an analogy between myth making and dreaming. This analogy is much further developed by psychoanalysis. Freud blazes a path with the following words: "The research into these concepts of folk psychology [myths, sagas, fairy stories] is at present not by any means concluded, but it is apparent everywhere from myths, for instance, that they correspond to the displaced residues of wish phantasies of entire nations, the dreams of ages of young humanity." (Samml. kl. Lehr. II, p. 205.) It will be shown later that fairy stories and myths can actually be subjected to the same psychologic interpretation as dreams, that for the most part they rest on the same psychological motives (suppressed wishes, that are common to all men) and that they show a similar structure to that of dreams.

Abraham (Traum und Mythus)[1] has gone farther in developing the parallelism of dream and myth. For him the myth is the dream of a people and a dream is the myth of the individual. He says, e.g., "The dream is (according to Freud) a piece

[1] See Translations in the Nervous and Mental Disease Monograph Series for this and the other studies cited in this section.

of superseded infantile, mental life " and " the myth
is a piece of superseded infantile, mental life of a
people "; also, " The dream then, is the myth of the
individual." Rank conceives the myths as images
intermediate between collective dreams and collective poems. " For as in the individual the dream
or poem is destined to draw off unconscious emotions
that are repressed in the course of the evolution of
civilization, so in mythical or religious phantasies a
whole people liberates itself for the maintenance of
its psychic soundness from those primal impulses
that are refractory to culture (titanic), while at the
same time it creates, as it were, a collective symptom
for taking up all repressed emotion." (Inz-Mot.,
p. 277. Cf. also Kunstl., p. 36.)

A definite group of such repressed primal impulses
is given a prominent place by psychoanalysis. I
refer to the so-called Œdipus complex that plays
an important rôle in the dream life as also in myth
and apparently, also in creative poetry. The fables
(sagas, dramas) of Œdipus, who slays his father
and marries his mother are well known. According to the observations of psychoanalysis there is a
bit of Œdipus in every one of us. [These Œdipus
elements in us can — as I must observe after reading Imago, January, 1913 — be called " titanic "
in the narrower sense, following the lead of Lorenz.
They contain the motive for the separation of the
child from the parents.] The related conflicts, that
in their entirety constitute the Œdipus complex (almost always unconscious, because actively re-

pressed) arise in the disturbance of the relation to
the parents which every child goes through more
or less in its first (and very early) sexual emotions.
" If king Œdipus can deeply affect modern mankind no less than the contemporary Greeks, the explanation can lie only in the fact that the effect of
the Greek tragedy does not depend on the antithesis
between fate and the human will, but in the peculiarity of the material in which this antithesis is
developed.   There must be a voice in our inner life
which is ready to recognize the compelling power of
fate in the case of Œdipus, while we reject as arbitrary the situations in the Ahnfrau or other destiny
tragedies.   And such an element is indeed contained
in the history of king Œdipus.   His fate touches
us only because it might have been ours, because the
oracle hung the same curse over us before our birth
as over him.   For us all, probably, it is ordained
that we should direct our first sexual feelings towards our mothers, the first hate and wish for violence against our fathers.   Our dreams convince us
of that.   King Œdipus, who has slain his father
Laius and married his mother Jocasta, is only the
wish-fulfillment of our childhood.   But more fortunate than he, we have been able, unless we have
become psychoneurotic, to dissociate our sexual feelings from our mothers and forget our jealousy of
our fathers.   From the person in whom that childish wish has been fulfilled we recoil with the entire
force of the repressions, that these wishes have since
that time suffered in our inner soul.   While the poet

in his probing brings to light the guilt of Œdipus,
he calls to our attention our own inner life, in which
that impulse, though repressed, is always present.
The antithesis with which the chorus leaves us

> See, that is Œdipus,
> Who solved the great riddle and was peerless in power,
> Whose fortune the townspeople all extolled and envied.
> See into what a terrible flood of mishap he has sunk.

This admonition hits us and our pride, we who have
become in our own estimation, since the years of
childhood, so wise and so mighty.   Like Œdipus,
we live in ignorance of the wishes that are so offensive to morality, which nature has forced upon us,
and after their disclosure we should all like to turn
away our gaze from the scenes of our childhood."
(Freud, Trdtg., p. 190 f.)

Believing that I have by this time sufficiently prepared the reader who was unfamiliar with psychoanalysis for the psychoanalytic part of my investigation, I will dispense with further time-consuming
explanations.

# II
# ANALYTIC PART

## Section I

# PSYCHOANALYTIC INTERPRETATION
# OF THE PARABLE

ALTHOUGH we know that the parable was written by a follower of the hermetic art, and apparently for the purpose of instruction, we shall proceed, with due consideration, to pass over the hermetic content of the narrative, which will later be investigated, and regard it only as a play of free fantasy. We shall endeavor to apply to the parable knowledge gained from the psychoanalytic interpretation of dreams, and we shall find that the parable, as a creation of the imagination, shows at the very foundations the same structure as dreams. I repeat emphatically that in this research, in being guided merely by the psychoanalytic point of view, we are for the time being proceeding in a decidedly one-sided manner.

In the interpretation of the parable we cannot apply the original method of psychoanalysis. This consists in having a series of seances with the dreamer in order to evoke the free associations. The dreamer of the parable — or rather the author — has long ago departed this life. We are obliged then to give up the preparatory process and stick to the methods derived from them.    There are three such methods.

The first is the comparison with typical dream images.    It has been shown that in the dreams of all individuals certain phases and types continually recur, and in its symbolism have a far reaching general validity, because they are manifestly built on universal human emotions.    Their imaginative expression is created according to a psychical law which remains fairly unaffected by individual differences.

The second is the parallel from folk psychology. The inner affinity of dream and myth implies that for the interpretation of individual creations of fancy, parallels can profitably be drawn from the productions of the popular imagination and vice versa.

The third is the conclusions from the peculiarities of structure of the dream (myth, fairy tale) itself. In dreams and still more significantly in the more widely cast works of the imagination creating in a dream-like manner, as e.g., in myths and fairy tales, one generally finds motives that are several times repeated in similar stories even though with variations and with different degrees of distinctness. [Let this not be misunderstood.  I do not wish to revive the exploded notion that myths are merely the play of a fancy that requires occupation.  My position on the interpretation of myths will be explained in Part I. of the synthetic part.]   It is then possible by the comparison of individual in-

stances of a motive, to conclude concerning its true
character, inasmuch as one, as it were, completes in
accordance with their original tendency the lines of
increasing distinctness in the different examples, and
thus — to continue the geometric metaphor — one
obtains in their prolongations a point of intersection
in which can be recognized the goal of the process
toward which the dream strives, a goal, however,
that is not found in the dream itself but only in the
interpretation.

We shall employ the three methods of interpretation conjointly.  After all we shall proceed exactly as psychoanalysis does in interpretation of folk-
lore.  For in this also there are no living authors
that we can call and question.  We have succeeded
well enough, however, with the derived methods.
The lack of an actual living person will be compen-
sated for in a certain sense by the ever living folk
spirit and the infinite series of its manifestations
(folk-lore, etc.).  The results of this research will
help us naturally in the examination of our parable,
except in so far as I must treat some of the conclusions of psychoanalysis with reserve as problematic.

Let us now turn to the parable.  Let us follow
the author, or as I shall call him, the wanderer, into
his forest, where he meets his extraordinary adventures.

I have just used a figure, " Let us follow him into
his forest."  This is worthy of notice.  I mean, of
course, that we betake ourselves into his world of
imagination and live through his dreams with him.

We leave the paths of everyday life, in order to rove in the jungle of phantasy. If we remember rightly, the wanderer used the same metaphor at the beginning of his narrative. He comes upon a thicket in the woods, loses the usual path. . . . He, too, speaks figuratively. Have we almost unaware, in making his symbolism our own, partially drawn away the veil from his mystery? It is a fact confirmed by many observations [Cf. my works on threshold symbolism — Schwellensymbolik, Jahrb. ps. F. III, p. 621 ff., IV, p. 675 ff.] that in hypnagogic hallucinations (dreamy images before going to sleep), besides all kinds of thought material, the state of going to sleep also portrays itself in exactly the same way that in the close of a dream or hypnotic illusions on awakening, the act of awakening is pictorially presented. The symbolism of awakening brings indeed pictures of leave taking, departing, opening of a door, sinking, going free out of a dark surrounding, coming home, etc. The pictures for going to sleep are sinking, entering into a room, a garden or a dark forest.

The fairy story also used the same forest symbol. Whether on sinking into sleep I have the sensation of going into a dark forest or whether the hero of the story goes into a forest (which to be sure has still other interpretations), or whether the wanderer in the parable gets into a tangle of underbrush, all amounts to the same thing; it is always the introduction into a life of phantasy, the entrance into the theater of the dream. The wanderer, if he had not chosen for his fairy tale the first person, could have begun as follows: There was once a king whose greatest joy was in the chase. Once as he was drawn with his companions into a great forest, and was pursuing a fleet stag, he was separated from his followers, and went still further from the familiar paths, so that finally he had to admit that he had lost his way. Then he went farther and farther into the woods until he saw far off a house. . . .

The wanderer comes through the woods to the Pratum felicitatis, the Meadow of Felicity, and there his adventures begin. Here, too, our symbolism is maintained; by sleeping or the transition to revery we get into the dream and fairy tale realm, a land to which the fulfillment of our keenest wishes beckons us. The realm of fairy tales is indeed — and the psychoanalyst can confirm this statement — a Pratum felicitatis, in spite of all dangers and accidents which we have there to undergo.

The dream play begins and the interpretation, easy till now, becomes more difficult. We shall hardly be able to proceed in strictly chronological order. The understanding of the several phases of the narrative does not follow the sequence of their events. Let us take it as it comes.

The wanderer becomes acquainted with the inhabitants of the Pratum felicitatis, who are discussing learned topics, he becomes involved in the scientific dispute, and is subjected to a severe test in order to be admitted to the company. The admission thus does not occur without trouble but rather

a great obstacle is placed in the way of it.   The
wanderer tells us that his examiners hauled him
over the coals, an allegorical metaphor, taken possibly from the ordeal by fire.   In these difficulties the
attaining of the end meets us in the first instance in
a series of analogous events, where the wanderer
sees himself hindered in his activities in a more or
less painful, and often even a dangerous manner.
After a phase marked by anxiety the adventure turns
out uniformly well and some progress is made after
the obstruction at the beginning.   As a first intimation of the coming experiences we may take up
the obstacles in the path in the first section of the
parable, which are successfully removed, inasmuch
as the wanderer soon after reaches the lovely region
(Sec. 3).

The psychology of dreams has shown that obstacles in the dream correspond to conflicts of will
on the part of the dreamer, which is exactly as in
the morbid restraint of neurotics.   Anxiety develops
when a suppressed impulse wishes to gratify itself,
to which impulse another will, something determined
by our culture, is opposed prohibitively.   Obstructed
satisfaction creates anxiety instead of pleasure.
Anxiety may then be called also a libido with a negative sign.   Only when the impulse in question knows
how to break through without the painful conflict,
can it attain pleasure — which is the psychic (not
indeed the biologic) tendency of every impulse
emanating from the depths of the soul.   The degrees of the pleasure that thus exists in the soul may

be very different, even vanishingly small, a state of affairs occurring if the wish fulfilling experience has through overgrowth of symbolism lost almost all of its original form. If we follow the appearances of the obstruction motive in the parable, and find the regular happy ending already mentioned, then we can maintain it as a characteristic of the phantasy product in question, that not only in its parts but also in the movement of the entire action, it shows a tendency from anxiety towards untroubled fulfillment of wishes.

As for the examination episode, to which we have now advanced in our progressive study of the narrative, we can now take up a frequently occurring dream type; the Examination Dream. Almost every one who has to pass severe examinations, experiences even at subsequent times when the high school or university examinations are far in the past, distressing dreams filled with the anxiety that precedes an examination. Freud (Trdtg., p. 196 ff.) clearly says that this kind of dream is especially the indelible memories of the punishments which we have suffered in childhood for misdeeds and which make themselves felt again in our innermost souls at the critical periods of our studies, at the *Dies irae, dies illa* of the severe examinations. After we have ceased to be pupils it is no longer as at first the parents and governesses or later the teachers that take care of our punishment. The inexorable causal nexus of life has taken over our further education, and now we dream of the preliminaries or finals; whenever we expect that the result will chastise us, because we have not done our duty, or done something incorrectly, or whenever we feel the pressure of responsibility. Stekel's experience is also to be noticed, confirmed by the practice of other psychoanalysts, that graduation dreams frequently occur if a test of sexual power is at hand. The double sense of the word *matura* (= ripe) (that may also mean sexual maturity) may also come to mind as the verbal connecting link for the association. In general the examination dreams may be the expression of an anxiety about not doing well or not being able to do well; in particular they are an expression of a fear of impotence. It should be noted here that not only in the former but in the latter case the fear has predominantly the force of a psychic obstruction.

For the interpretation of the examination scene also, we note the fairy tale motif so frequently appearing of a hard-won prize, i.e., any story in which a king or a potentate proposes a riddle or a task for the hero. If the hero solves or accomplishes it, he generally wins, besides other precious possessions, a woman or a princess, whom he marries. In the case of a heroine the prize is a beautiful prince. The motif of the hard-won prize matches the later appearance of obstacles in the Parable. The nature of the prize is, for the present at any rate, a matter of indifference.

A second edition of the examination scene meets us in the 6th section as the battle with the lion.

The advance from the anxiety phase to the fulfillment phase appears clearly and the emotions of the
wanderer are more strongly worked out.   The difficulty at the beginning is indicated in the preceding
conversation where no one will advise him how he
is to begin with the beast, but all hold out guidance
for a later time when he shall have once bound the
lion.   The beginning of the fight causes the wanderer much trouble.   He " is amazed at his own
temerity," would gladly turn back from his project,
and he can hardly restrain his tears for fear."   He
fortifies himself, however, develops brilliant abilities and comes off victor in the fray.   A gratification derived from his own ability is unmistakable.
The scene, as well as a variation of the preceding
examination, adds to it some essentially new details.
The displacement of the early opponents (i.e., the
examining elders) by another (the lion) is not really
new.   It is a mere compensation, although, as we
shall see later, a very instructive one.   Entirely new
is the result of the battle.   After killing the lion the
victor brings to light white bones and red blood
from his body.   Note the antithesis, white and red.
It will occur again.   If we think of saga and fairy
lore parallels, the dragon fight naturally comes to
mind.   The victorious hero has to free a maiden
who languishes in the possession of an ogre.   The
anatomizing of the dead lion finds numerous analogies in those myths and fairy tales in which dismemberment of the body appears.   It will be dealt with
fully later on.

As the next obstacle in the parable we meet the difficult advance on the wall.    (Para. 7 and 8.) We have here again an obstruction to progress in the narrower sense as in Sec. 1, but with several additions.    The wall, itself a type of embarrassment, reaches up to the clouds.    Whoever goes up so high may fall far.    The way on top is "not a foot in width" and an iron hand rail occupies some of that space.    The walking is therefore uncomfortable and dangerous.    The railing running in the middle of it divides the path and so produces two paths, a right and a left.    The right path is the more difficult. Who would not in this situation think of Hercules at the cross roads?    The conception of right and left as right and wrong, good and bad, is familiar in mythical and religious symbolism.    That the right path is the narrower [Matth. VII, 13, 14] or more full of thorns is naturally comprehensible. In dreams the right-left symbolism is typical.    It has here a meaning similar to its use in religion, probably however, with the difference that it is used principally with reference to sexual excitements of such a character that the right signifies a permitted (i.e., experienced by the dreamer as permissible), the left, an illicit sexual pleasure.    Accordingly it is, e.g., characteristic in the dream about strawberry picking in the preceding part of the book, that the valley, sought by the dreamer and the boy, " in order to pick strawberries there " turns off to the left from the road, not to the right.    The sexual act with a boy appears even in dreams as something illicit, indecent, forbidden. In the parable the wanderer goes from right to left, gets into difficulties by doing so, but knows, as always, how to withdraw successfully.

From the wall the wanderer comes to a rose tree, from which he breaks off white and red roses. Notice the white and red. The victory over the lion has yielded him white bones and red blood, the passing through the dangers on the wall now yields him white and red roses. The similarity in the latter case is particularly marked by his putting them in his hat.

Again in the course of the next sections (9–11) there are obstacles. There a wall is set up against the wanderer. On account of that he has, in order to gain entrance for the maidens into the company in the garden, to go a long way round. Arriving at the door, he finds it locked and is afraid that the people standing about will prevent his entrance or laugh at him. But the first difficulty is barely removed, by the magic opening of the first gate, when the now familiar change from the anxiety phase to the fulfillment phase occurs. The wanderer traverses the corridor without trouble but his eyes glance ahead of him and he sees through the still closed door, as if it were glass, into the garden. What result has this success over the difficulties yielded him? Where is the usual white and red reward? We do not have to look long. In Sec. 11 it is recorded, " When I had passed beyond the little garden [in the center of the larger garden] and was going to the place where I was to help the maidens, behold, I was aware that instead of the wall, a low hurdle stood there, and there went by the rose garden, the most beautiful maiden arrayed in white satin with the most stately youth who was in scarlet, each giving arm to the other, and carrying in their hands many fragrant roses. 'This, my dearest bridegroom,' said she, ' has helped me over and we are now going out of this lovely garden into our chamber to enjoy the pleasures of love.' " Here the parallel with the fairy tale is complete, and reveals the characteristic of the prize that rewards him. The red and the white reveal themselves as man and woman, and the last aim is, as the just quoted passage clearly shows, and the further course of the narrative fully indicates, the sexual union of both. Even the rest of the fairy tale prizes are not lacking — kingdoms, riches, happiness. And if they are not dead they are still living. . . . The narrative has yielded a complete fulfillment of wishes; the longing for love and power has attained its end. That the wanderer does not experience the acquired happiness immediately in his own person, but that the representation of happy love is in the most illustrative manner developed in the union of two other persons, is naturally a peculiarity of the narration. It is found often enough in dreams. The ego of the dreamer is in such a case replaced by a " split-off " person, through whom the dream evokes its dramatic pageantry. It is as if the parable tried to say the hero has won his happy love through struggle; two are, however, needful for love, a man and a woman, so let us quickly create a pair. Apart from the fact that the reward must evidently fall to the hero who has won it, the identity of the wanderer with the king in the parable is abundantly demonstrated, even if somewhat paraphrased. The secret of the dramatizing craft of the narrative is most clearly exposed in the conclusion of Sec. 11, where the elders, with the letter of the faculty in their hands, reveal to the wanderer that he must marry the woman he has taken, which he furthermore cheerfully promises them to do.

So far all would be regular and we might think, on superficial examination, that the psychoanalytic solution of the parable was ended. How far from being the case! We have interpreted only the upper stratum and will see a problem show itself that invites us to press on into the deeper layers of the phantasy fabric before us.

We have noticed that in the parable much, even the most important, is communicated only by symbols and by means of allusions. Its previously ascertained latent content [corresponding to the latent dream thoughts] will in the manifest form be transcribed in different and gradually diminishing disguises. Also a displacement (dream displacement) has taken place. Now the dream or the imagination working in dreams does nothing without purpose and even though according to its nature (out of " regard for presentability ") it has to favor the visual in all cases, the tendency toward the pictorial

does not explain such a systematic series of disguises
and such a determinate tendency as that just observed by us. The representation of the union of
man and woman is strikingly paraphrased. First
as blood and bones — a type of intimate vital connection; they belong to *one* body, just as two levers
are one and as later the bridal pair also melt into
one body. Then as two kinds of roses that bloom
on one bush. The wanderer breaks the rose as the
boy does the wild rose maiden. And hardly is the
veil of the previous disguise lifted, hardly have we
learned that the wanderer has taken a woman
(Sec. 11), when the affair is again hushed just as
it is about to be dramatized (cf. Sec. 12), so that
apparently another enjoys the pleasures of love.
This consequent concealment must have a reason.
Let us not forget the striking obstacles which the
wanderer experiences again and again and which
we have not yet thoroughly examined. The symbolism of the dream tells us that such obstacles correspond to conflicts of the will. What kind of inner
resistance may it be that checks the wanderer at
every step on his way to happy love? We suspect
that the examinations have an ethical flavor. This
appears to some extent in the right-left symbolism;
then in the experience at the mill, which we have not
yet studied, where the wanderer has to pass over a
very narrow plank, the ethical symbolism of which
will be discussed later; and in the striking feeling
of responsibility which the wanderer has for the
actions of the bridal pair in the crystal prison, which

gives us the impression that he had a bad conscience.
Altogether we cannot doubt that the dream — the
parable — has endeavored, because of the censor,
to disguise the sexual experiences of the wanderer.
We can be quite certain that it will be said that the
sexual as such will be forbidden by the censor.    That
is, however, not the case.    The account is outspoken
enough, and not the least prudish; the bridal pair
embrace each other naked, penetrate each other and
dissolve in love, melt in rapture and pain.    Who
could ask more?    Therefore the sexual act itself
could not have been offensive to the censor.    The
whole machinery of scrupulousness, concealment
and deterrent objects, which stand like dreadful
watchmen before the doors of forbidden rooms, cannot on the other hand be causeless.    So the question
arises:    What is it that the dream censor in the most
varied forms [lion, dangerous paths, etc.] has so
sternly vetoed?

In the strawberry dream related in the preceding
section, we have seen that a paraphrase of the latent
dream content appears at the moment when a form
of sexual intercourse, forbidden to the dreamer
by the dream censor, was to be consummated.
(Homosexual intercourse.)    Most probably in the
parable also there is some form of sexuality rejected
by the censor.    What may it be?    Nothing indicates a homosexual desire.    We shall have to look
for another erotic tendency that departs from the
normal.    From several indications we might settle
upon exhibitionism.    This is, as are almost all abnor-

mal erotic tendencies, also an element of our normal psychosexual constitution, but it is, if occurring too prominently, a perversity against which the censor directs his attacks. The incidents of the parable that indicate exhibitionism are those where the wanderer sees, through locked doors (Sec. 10) or walls (Sec. 11), objects that can be interpreted as sexual symbols. The miraculous sight corresponds to a transferred wish fulfillment. The supposition that exhibitionism is the forbidden erotic impulse element that we were looking for is, however, groundless, if we recollect that these very elements appear most openly in the parable. In Sec. 14 the wanderer has the freest opportunity to do as he likes. Still the question arises, what is the prohibited tendency? No very great constructive ability is required to deduce the answer. The wording of the parable itself furnishes the information. In Sec. 14 we read, " Now I do not know what sin these two have committed except that although they were brother and sister they were so united in love that they could not again be separated and so, as it were, required to be punished for incest." And in another passage (Sec. 13), " After our bridegroom . . . with his dearest bride . . . came to the age of marriage, they both copulated at once and I wondered not a little that this maiden, that yet was supposed to be the bridegroom's mother, was still so young."

The sexual propensity forbidden by the censor is incest. That it can be mentioned in the parable in spite of the censor is accounted for by the exceedingly clever and unsuspected bringing about of the suggestion. Dreams are very adroit in this respect, and the same cleverness (apparently unconscious on the part of the author) is found in the parable, which is in every way analogous to the dream. Incest can be explicitly mentioned, because it is attributed to persons that have apparently nothing to do with the wanderer. That the king in the crystal prison is none other than the wanderer himself, we indeed know, thanks to our critical analysis. The dreamer of the dream does not know it. For him the king is a different person, who is alone responsible for his actions; although in spite of the clear disguise, some feeling of responsibility still overshadows the wanderer, a peculiar feeling that has struck us before, and now is explained.

Later we shall see that from the beginning of the parable, incest symbols are in evidence. Darkly hinted at first they are later somewhat more transparent, and in the very moment when they remove the last veil and attain a significance intolerable for the censor, exactly at that psychologic moment the forbidden action is transferred to the other, apparently strange, person.

A similar process, of course, is the change of situation in the strawberry dream at the exact moment when the affair begins to seem unpleasant to the dreamer. This becoming unpleasant can be beautifully followed out in the parable. The critical transition is found exactly in one of those places where the representation appears most confused.

It is in this way that the weakest points of the dream
surface are usually constituted. Those are the
places where the outer covering is threadbare and
exposes a nakedness to the view of the analyzer.

The critical phase of the parable begins in the
11th section. The elders consult over a letter from
the faculty. The wanderer notices that the contents
concern him and asks, " Gentlemen, does it have to
do with me? " They answer, " Yes, you must
marry your woman that you have recently taken."
Wanderer: " That is no trouble; for I was, so to
speak, born [how subtle!] with her and brought up
from childhood with her." Now the secret of the
incest is almost divulged. But it is at once ef-
fectually retracted. In Sec. 12 we read, " So my
previous trouble and toil fell upon me and I be-
thought myself that from strange causes [these
strange causes are the dream censor who, ruling in
the unconscious, effects the displacements that fol-
low], it cannot concern me but another that is well
known to me [in truth a well-known other]. Then
I see our bridegroom with his bride in the previous
attire going to that place ready and prepared for
copulation and I was highly delighted with it. For
I was in great anxiety lest the affair should concern
me." The anxiety is quite comprehensible. It is
just on account of its appearance that the displacement from the wanderer to the other person takes
place. Further in Sec. 13: " Now after . . . our
bridegroom . . . with his dearest bride . . . came
to the age of marriage [The aim with which the

censor performs his duties and effects the dream displacement is, says Freud (Trdtg., p. 193), " to prevent the development of anxiety or other form of painful affect.] they both copulated . . . and I wondered not a little that this maiden, that was supposed to be actually the *mother* of the bridegroom, was still so young. . . ." Now when the transfer has taken place, the thought of its being the mother is hazarded; whereas formerly a mere suggestion of a sister had been offered. Section 14 explicitly mentions incest and even arranges the punishment of the guilt. In this form the matter can, of course, be contemplated without troubling the conscience or being further represented pictorially.

The sister, alternating in the narrative with the mother, is only a preliminary to the latter. As we find that the Œdipus complex [Rather an attenuation, which occurs frequently, not merely in dream psychology, but also in modern mythology.] is revived in the parable, let us bring the latter into still closer relation with the fairy tales and myths to which we have compared it. The woman sought and battled for by the hero appears, in its deeper psychological meaning, always to be the mother. The significance of the incest motive has been discovered on the one hand by the psychoanalysts (in particular Rank, who has worked over extensive material), on the other by the investigators of myths. That many modern mythologists lay most stress in this discovery upon the astral or meteorological content and do not draw the psychological conclusions is another matter that will be discussed later.   But in passing it may be noted that the correspondence in the discovered material (motives) is the more remarkable as it resulted from working in the direction of quite different purposes.

It is now time to examine the details of the parable in conformity with the main theme just stated and come to a definite interpretation.   Henceforward we may keep to a chronological order.

The threshold symbolism in the beginning of the parable has already been given, also the obstacles that are indicative of a psychic conflict.   We might rest satisfied with that, yet a more complete interpretation is quite possible, in which particular images are shown to be overdetermined.   The way is narrow, overgrown with bushes, and leads to the Pratum felicitatis.   That, according to a typical dream symbolism, is also a part of the female body. The obstacles in the way we recognize as a recoil from or impediment to incest; so it is evident that a definite female body, namely that of the mother, is meant.   The penetration leads to the Pratum felicitatis, to blissful enjoyment.   In fairy lore the sojourn in the forest generally signifies death or the life in the underworld.   Wilhelm Muller, for example, writes, " As symbols of similar significance we have the transformation into swans or other birds, into flowers, the exposure in the forest, the life in the glass mountain, in a castle, in the woods. . . . All imply death and life in the underworld." The underworld is, when regarded mythologically, not only the land where the dead go, but also whence the living have come; thence for the individual, and in particular for our wanderer, the uterus of the mother. It is significant that the wanderer, as he strolls along, ponders over the fall of our first parents and laments it. The fall of the parents was a sexual sin. That it was incest besides, will be considered later. The son who sees in his father his rival for his mother is sorry that the parents belong to each other. A sexual offense (incest) caused the loss of paradise. The wanderer enters the paradise, the Pratum felicitatis. [Garden of Joy, Garden of Peace, Mountain of Joy, etc., are names of paradise. Now it is particularly noteworthy that the same words can signify the beloved. (Grimm, D. Mythol., II, pp. 684 ff., Chap. XXV, 781 f.)] The path thither is not too rough for him (Sec. 2).

In Sec. 3 the wanderer enters his paradise (incest). He finds in the father an obstacle to his relation with the mother. The elders (splitting of the person of the father) will not admit him, forbid his entrance into the college. He himself, the youth is already among them. The younger man, whose name he knows without seeing his face, is himself. He puts himself in the place of his father. (The other young man with the black pointed beard may be an allusion to a quite definite person, intended for a small circle of readers of the parable, contemporaries of the author. Either the devil or death may be meant, yet I cannot substantiate this conjecture.)

In the fourth section the examinations begin. First the examination in the narrower sense of the word. The paternal atmosphere of every examination has already been emphasized in the passage from Freud quoted above. Every examination, every exercise is associated with early impressions of parental commands and punishments. Later (in the treatment of the lion) the wanderer will turn out to be the questioner, whereas now the elders are the questioners. In the relation between parent and child questions play a part that is important from a psychological point of view. Amazingly early the curiosity of the child turns toward sexual matters. His desire to know things is centered about the question as to where babies come from. The uncommunicativeness of the parents causes a temporary suppression of the great question, which does not, however, cease to arouse his intense desire for explanation. The dodging of the issue produces further a characteristic loss of trust on the part of the child, an ironic questioning, or a feeling that he knows better. The knowing better than the questioning father we see in the wanderer. The tables are turned. Instead of the child desiring (sexual) explanation from the parents, the father must learn from the child (fulfillment of the wish to be himself the father, as above). The elders are acquainted only with figurative language (" Similitudines," " Figmenta," etc.) ; but the wanderer is well informed in practical life, in experience

he is an adept.    As a fact, parents in their indefi-
niteness about the question, Where do babies come
from? give a figurative answer (however appropriate it may be as a figure of speech) in saying that
the stork brings them, while the child expects clear
information (from experience).    On the propriety
of the picturesque information that the stork brings
the babies out of the water we may note incidentally
the following observations of Kleiñpaul.    The fountain is the mother's womb, and the red-legged stork
that brings the babies is none other than a humorous
figure for the organ (phallus) with the long neck
like a goose or a stork, that actually gets the little
babies out of the mother's body.    We understand
also that the stork has bitten Mamma in the " leg."

We have become acquainted above with the fear
of impotence as one significance of the anxiety about
examinations.    Psychosexual obstructions cause impotence.    The incest scruple is such an obstruction.

According to Laistner we can conceive the painful
examination as a question torture — a typical experience of the hero in countless myths.    Laistner,
starting from this central motive, traces the majority
of myths back to the incubus dream.    The solution
of the tormenting riddle, the magic word that ban-
ishes the ghost, is the cry of awakening, by which
the sleeper is freed from the oppressing dream, the
incubus.    The prototype of the tormenting riddle
propounder is, according to Laistner, the Sphinx.
Sphinx, dragon, giants, man eaters, etc., are analo-

gous figures in myths. They are what afflict the heroes, and what he has to battle with. The corresponding figure in our parable is the lion.

Although the wanderer has brilliantly stood the test, the elders (Sec. 5) do not admit him into their college (the motive of denial recurs later); but enter him for the battle with the lion. This is surely a personification of the same obstructions as the elders themselves. In them we have, so to speak, before us the dragon (to be subdued) in a plural form. Analogous multiplying of the dragon is found, for example, in Stucken [in the astral myth]. Typical dragon fighters are Jason, Joshua, Samson, Indra; and their dragon enemies are multitudes like the armed men from the sowing of the dragon's teeth by Jason, the Amorites for Joshua, the Philistines in the case of Samson, the Dasas in that of Indra. We know that for the wanderer the assemblage of elders is to be conceived chiefly as the father, and the same is true now of the lion (king of animals, royal beast, also in hermetic sense) who has as lion been already appropriated to the father symbol. Kaiser, king, giants, etc., are wont in dreams to represent the father. Accordingly large animals, especially wild beasts or beasts of prey, are accustomed to appear in dreams with this significance.

Stekel [Spr. Tr.] contributes the following dream of the patient Omicron: " I was at home. My family had preserved a dead bear. His head was of wood and out of his belly grew a mighty tree, which looked very old.  Around the animal's neck was a chain.  I pulled at it, and afterwards was afraid that I had possibly choked him, in spite of the fact that he was long dead."

And the following interpretation of it derived through analysis:  " The bear is a growler, i.e., his father, who has told him many a lie about the genesis of babies.  He reviles him for it.  He was a blockhead, he had a wooden head.  The mighty tree is the phallus.  The chain is marriage.  He was a henpecked husband, a tamed bear.  Mother held him by the chain.  This chain (the bond of marriage) Omicron desired to sunder.  (Incest thoughts.)  When the father died Omicron held his hand over his father's mouth to find out whether he was still breathing.  Then he was pursued by compulsive ideas, that he had killed his father.  In dreams the same reproaches appear.  We realize how powerful his murder impulses were.  His reproaches are justified.  For he had countless death wishes that were centered about that most precious life."

A girl not yet six years old told her mother the following dream:

" We went together, there we saw a camel on a rock, and you climbed up the cliff.  The camel wanted to keep slobbering you, but you wouldn't let him, and said, ' I'd like to do it, but if you are like this, I won't do it.' "

After the telling of the dream the mother asked the girl if she could imagine what the camel signi-

fied in the dream, and she immediately replied:
" Papa, because he has to drag along and worry himself like a camel.  You know, Mamma, when he
wants to slobber you it is as if he said to you in
camel talk, ' Please play with me.  I will marry
you; I won't let you go away.'  The rocks on which
you are were steep, the path was quite clear, but the
railing was very dirty and there was a deep abyss,
and a man slipped over the railing into the abyss.
I don't know whether it was Uncle or Papa."

Stekel remarks on this: " The neurotic child
understands the whole conflict of the parents.  The
mother refused the father coitus.  In this she will
not ' play ' with the camel.  The camel wants to
' marry ' her.  It is quite puzzling how the child
knows that Mamma has long entertained thoughts
of  separation. . . . Children  evidently  observe
much more sharply and exactly than we have yet suspected.  The conclusion of the dream is a quite
transparent symbolism of coitus.  But the dream
thoughts go deeper yet.  A man falls into an abyss.
The father goes on little mountain expeditions.
Does the child wish that the father may fall?  The
father treats the child badly and occasionally strikes
her unjustly.  At all events it is to be noted that
the little puss says to her mother, ' Mamma, isn't
it true that when Papa dies you will marry Dr.
Stekel? '  Another time she chattered, ' You know,
Mamma, Dr. N. is nicer than Papa; he would suit
you much better.'  Also the antithesis of clean and

dirty, that later plays such an important part in the psychic life of neurotics, is here indicated."

Not only the camel but also the railing and the abyss are interesting in relation to Sec. 7 and 8 of the parable, where occurs the perilous wall with the railing. People fall down there. There is evidently here an intimate primitive symbolism (for the child also). But I will not anticipate.

It is not necessary to add anything to the bear dream. It is quite clear. Only one point must be noticed, that the subsequent concern about the dead is to be met in the parable, though not on the wanderer's part but on that of the elders who desire the reviving of the lion.

The wanderer describes the lion (Sec. 6) as " old, fierce and large." (The growling bear of the dream.) The glance of his eye is the impressively reproachful look of the father.

The wanderer conquers the lion and " dissects " him. Red blood, white bones, come to view, male and female; the appearance of the two elements is, at any rate overdetermined in meaning as it signifies on the one hand the separation of a pair, father and mother, originally united as one body; and on the other hand the liberation of sexuality in the mind of the wanderer (winning of the mother or of the dragon-guarded maiden).

We ought not to explain the figures of the lion and the elders as " the father." Such exalted figures are usually condensations or composite persons.

The elders are not merely the father, but also the old, or the older ones = parents in general, in so far as they are severe and unapproachable. Apparently the mother also will prove unapproachable if the adult son desires her as a wife. [The male child, on the other hand, frequently has erotic experiences with the mother. The parents connive at these, because they do not understand the significance even of their own caresses. They generally do not know how to fix the limits between moderation and excess.] The wanderer has no luck with blandishments in the case of the lion. He begins indeed to fondle him (cf. Sec. 6), but the lion looks at him formidably with his bright, shining eyes. He is not obliging; the wanderer has to struggle with him. Offering violence to the mother often appears in myths. We shall have an example of this later. It is characteristic that the wanderer is amazed at his own audacity.

Dragon fighting, dismembering, incest, separation of parents, and still other motives have an intimate connection in mythology. I refer to the comparison of motives collected by Stucken from an imposing array of material. [I quote an excerpt from it at the end of this volume, Note A.] The motive of dismemberment has great significance for the subsequent working out of my theme, so I must for that reason delay a little longer at this point.

The parts resulting from the dismemberment have a sexual or procreative value. That is evident from the analysis of the parable, even without the support of mythological parallels.   None the less let it be noticed that many cosmogonies assign the origin of the universe or at least the world or its life to the disintegrated parts of the body of a great animal or giant.   In the younger Edda the dismemberment of the giant Ymir is recounted.

" From Ymir's flesh was the earth created,
From his sweat the sea,
From his skeleton the mountains, the trees from his hair,
From his skull the heavens,
From his eyebrows kindly Äses made
Mitgard, the son of man.
But from his brain were created all the ill-tempered clouds."

The Iranian myth has an ancestor bull, Abudad. " From his left side goes Goschorum, his soul, and rises to the starry heavens; from his right side came forth Kajomorts (Gâyômard), the first man.   Of his seed the earth took a third, but the moon two thirds.   From his horns grew the fruits, from his nose, leeks, from his blood, grapes, from his tail, five and twenty kinds of grain.   From his purified seed two new bulls were formed, from which all animals are descended."   Just as in the Iranian myth the original being, Gâyômard, considered as human, and the ancestor bull belong together, so we find in the northern myth a cow Audhumla associated with Ymir.   Ymir is to be regarded as androgynous (man and woman), the primitive cow as only a doubling of his being.   The Iranian primitive bull ancestor also occurs as cow.   Compare white and red, male and female, in the body of the lion.

In the Indian Asvamedha the parts of the sacrificed steed correspond to the elements of the visible creation.   (Cf. Brhadaranyaka — Upanisad I, 1.) A primitive vedic cosmogony makes the world arise from the parts of the body of a giant.   (Rig-veda purusa-sukta.)

Just as from the dead primordial being the sacrificed bull, Mithra, sprouts life and vegetation, so in the dream of Omicron, a tree grows out of the belly of the dead bear.   In mythology many trees grow out of graves, that in some way reincarnate the creative or life principle of the dead.   It is an interesting fact that the world, or especially an improved new edition of the world, comes from the body of a dying being.   Some one kills this being and so causes   an   improved   creation.   (According   to Stucken, incidentally, all myths are creation myths.) This improvement is now identical, psychologically, with the above mentioned superior knowledge of the son (expressed in general terms, the present new generation as opposed to the ancestors).   The son does away with the father (the children overpower the ancestors), and creates, as it were out of the wreckage, an improved world.   So, beside the superior knowledge, a superior efficiency.   The primordial beings are destroyed but not so the creative power (phallus, tree, the red and the white).   It passes on to posterity (son) which uses it in turn.

Dismemberments in creation myths are not always multiple but sometimes dichotomous.   Thus in the Babylonian cosmogony Marduk splits the monster

Tiamat into two pieces, which henceforth become the upper and lower half of heaven. Winckler concludes that Tiamat is man-woman (primal pair). This brings us to the type of creation saga where the producer of the (improved) world separates the primal pair, his parents. The Chinese creation myth speaks of the archaic Chaos as an effervescing water, in which the two powers, Yang (heaven) and Yin (earth), the two primal ancestors, are mingled and united. Pwanku, an offshoot of these primal powers (son of the parents), separates them and thus they become manifest. In the Egyptian myth we read (in Maspero, Histoire des Peuples de l'Orient, Stucken, Astral Myth, p. 203) : " The earth and the heaven were in the beginning a pair of lovers lost in the Now who held each other in close embrace, the god below the goddess. Now on the day of creation a new god [son type], Shou, came out of the eternal waters, glided between them and seizing Nouit [the goddess] with his hands, lifted her at arms' length above his head. While the starry bust of the goddess was lengthened out in space, the head to the west, the loins to the east, and became the sky, her feet and her hands [as the four pillars of heaven] fell here and there on our earth." The young god or the son pushes his way between the parents, sunders their union, just as the dreamer Omicron would have liked to sunder the chain of the bear (the marriage bond of the parents). This case is quite as frequent a type in analytic psychology as in mythical cosmology. The child is actually an intruder, even if it does indirectly draw the bonds of marriage tighter.   Fundamentally regarded, the child appears as the rival of the father, who is no longer the only beloved one of his wife.   He must share the love with the new comer, to whom an even greater tenderness is shown.   Regarded from the standpoint of the growing son, the intrusion represents the Œdipus motive (with the incest wish).

The most outspoken and also a commonly occurring form of the mythological separation of the primal pair is the castration of the father by the son.   The motive is, according to all accounts, psychologically quite as comprehensible as the frequently substituted castration of the son by the father.   The latter is psychologically the necessary correlate of the first form.   The rivals, father and son, menace each other's sexual life.   That the castration motive works out that way with father and son (son-in-law if the daughter takes the place of the mother) is expressed either in so many words in the myth or through corresponding displacement types.

A clear case is the emasculation of Uranus by his son Kronos, who thereby prevents the further cohabitation of the primal parents.   [Archetype of the Titan motive in a narrow sense.]   Important for us is the fact that castration in myths is represented sometimes as the tearing out of a limb or by complete dismemberment.   (Stucken, Astral Myth, pp. 436, 443, 479, 638 ff.; Rank, Incest Motive, p. 311 ff.)

The Adam myth also contains the motive of the separated primal parents.    In Genesis we do not, of course, see the myth in its pure form.    It must first be rehabilitated.    Stucken accomplishes this in regarding Adam and Eve (Hawwa) as the original world-parent pair, and Jahwe Elohim as the separating son god.    By a comparison of Adam and Noah he incidentally arrives by analogical reasoning at an emasculation of Adam.    In connection with the " motive of the sleeping primal father," he observes later (Astral Myth, p. 224) that the emasculation (or the shameless deed, Ham with Noah) is executed while the primal father lies asleep. Thus, Kronos emasculates Uranus by night while he is sleeping with Gaia.    Stucken now shows that the sleep motive is contained in the 2d chapter of Genesis.    " And the Lord God caused a deep sleep to fall upon Adam, and he slept; and he took one of his ribs, and closed up the flesh instead thereof." (II, 21.)    According to Stucken the rib stands euphemistically for the organ of generation, which is cut off from Adam while he sleeps.

Rank works out another kind of rearrangement. He takes the creation of Eve from Adam as an inversion.    He refers to the ever recurring world-parent myths of savage peoples, in which the son begets upon the mother a new generation.    He cites after Frobenius a story from Joruba, Africa, where the son and daughter of the world parents marry and have a son, who falls in love with his mother. As she refuses to yield to his passion he follows and overpowers her. She immediately jumps up and runs away crying. The son follows her to soothe her, and when he catches her she falls sprawling on the earth, her body begins to swell, two streams of water spring from her breasts and her body falls in pieces. Fifteen gods spring from her disrupted body. [Motive of the mutilation of the maternal body. The dismembered lion also naturally contains this motive. From the mutilated body come male and female (red and white) children.] Rank supposes that the biblical account of the world parents serves as a mask for incest (and naturally at the same time the symbolic accomplishment of the incest). He continues, " It is needed only that the infantile birth theory [Birth from anus, navel, etc. The taking of the rib = birth process.] which ignores the sexual organs in woman and applies to both sexes, be raised in the child's thought to the next higher grade of knowledge, which ascribes to the woman alone the ability to bring children into the world by the opening of her body. In opposition to the biblical account we have the truly natural process, according to which Adam came out of the opened body of Eve. If by analogy with other traditions, we may take this as the original one, it is clear that Adam has sexual intercourse with his mother, and that the disguising of this shocking incest furnished the motive for the displacement of the saga and for the symbolic representation of its contents." The birth from the side of the body, from the navel, from the anus, etc., are among children

common theories of birth.    In myths analogous to
the biblical apple episode the man almost always
offers the apple to the woman.    The biblical account
is probably an inversion.    The apple is an apple of
love and an impregnation symbol.    Impregnation by
food is also an infantile procreation theory.    For
Rank, therefore, it is Adam who is guilty of separating the primal parents [Jahwe and Hawwa] and of
incest with the mother.    The contrast between the
two preceding conceptions of the Adam myth should
not be carried beyond limits.    That they can stand
side by side is the more conceivable because Genesis
itself is welded together from heterogeneous parts
and  different elaborations of the primal pair motive.
Displacements, inversions, and therefore apparent
contradictions must naturally lie in such a material.
Moreover, the interpretation depends not so much
on the narrative of the discovered motives as on the
motives themselves.    [On the interpretation of the
mythological motives cf. Lessmann, Aufg. u. Ziele,
p. 12.]

Let us return now to the motive of dismemberment.    One of the best known examples of dismemberment in mythology is that of Osiris.    Osiris and
Isis, the brother and sister, already violently in love
with each other in their mother's womb, as the myth
recounts, copulated with the result that Arueris was
born of the unborn.    So the two gods came into the
world as already married brother and sister.    Osiris
traversed the earth, bestowing benefits on mankind.
But he had a bad brother, full of jealousy and envy,

Typhon (Set), who would gladly have taken advantage of the absence of his brother to place himself on his throne. Isis, who ruled during the absence of Osiris, acted so vigorously and resolutely that all his evil designs were frustrated. Finally Osiris returned and Typhon, with a number of confederates (the number varies) and with the Ethiopian queen Aso, formed a conspiracy against the life of Osiris, and in feigned friendship arranged a banquet. He had, however, caused a splendid coffin to be made, and as they sat gayly at the feast, Typhon had it brought in, and offered to give it to the person whose body would fit it. He had secretly taken the measure of Osiris and had prepared the coffin accordingly. All tried it in turn. None fitted. Finally Osiris lay in it. Then Typhon and his confederates rushed up, closed it and threw it into the river, which carried it to the sea. (Creuzer, L., p. 259 ff.) For the killing of his brother Set, which happened according to the original version on account of desire for power, later tradition substitutes an unconscious incest which Osiris committed with his second sister, Nephthys, the wife of Set, a union from which sprang Anubis (the dog-headed god). Set and Nephthys are, according to H. Schneider, apparently no originally married brother and sister like Osiris and Isis, but may have been introduced by way of duplication, in order to account for the war between Osiris and his brother. With the help of Anubis, Isis finds the coffin, brings it back to Egypt, opens it in seclusion and gives way to her tender feelings and sorrow for him.  Thereupon she hides the coffin with the body in a thicket in the forest in a lonely place.  A hunt which the wild hunter Typhon arranges, discovers the coffin.  Typhon cuts the body into fourteen pieces.  Isis soon discovers the loss and searches in a papyrus canoe for the dismembered body of Osiris, traveling through all the seven mouths of the Nile, till she finally has found thirteen pieces.  Only one is lacking, the phallus, which had been carried out to sea and swallowed by a fish.  She put the pieces together and replaced the missing male member by another made of sycamore wood and set up the phallus for a memento (as a sanctuary).  With the help of her son Horus, who, according to later traditions, was begotten by Osiris after his death, Isis avenged the murder of her spouse and brother. Between Horus and Set, who originally were brothers themselves, there arises a bitter war, in which each tore from the other certain parts of the body as strength-giving amulets.  Set knocked an eye out of his opponent and swallowed it, but lost at the same time his own genitals (testicles), which in the original version were probably swallowed by Horus. Finally Set was overcome and compelled to give up Horus' eye, with the help of which Horus again revivified Osiris so that he could enter the kingdom of the dead as a ruler.

The dismemberment, with final loss of the phallus, will be clearly recognized as a castration.  The tearing out of the eye is similarly to be regarded as emas-

culation.   This motive is found as self-punishment
for incest, at the close of the Œdipus drama.   On
the dismemberment of Osiris as a castration, Rank
writes (Inz. Mot., p. 311): " The characteristic
phallus consecration of Isis shows us that her sorrow predominantly concerns the loss of the phallus, (and it also is expressed in the fact that according to a later version, she is none the less in a
mysterious manner impregnated by her emasculated
spouse), so on the other hand the conduct of the
cruel brother shows us that in the dismemberment he
was particularly interested in the phallus, since that
indeed was the only thing not to be found, and had
evidently been hidden with special precautionary
measures.   Indeed both motivations appear closely
united in a version cited by Jeremias (Babylonisches
in N. T., p. 721), according to which Anubis, the son
of the adulterous union of Osiris with his sister
Nephthys, found the phallus of Osiris, dismembered
by Typhon with 27 assistants, which Isis had hidden
in the coffin.   Only in this manner could the phallus
from which the new age originated, escape from
Typhon.   If this version clearly shows that Isis
originally had preserved in the casket the actual phallus of her husband and brother which had been made
incorruptible and not merely a wooden one, then on
the other hand the probability increases that the
story originally concerns emasculation alone because
of the various weakening and motivating attempts
that meet us in the motive of the dismemberment."
   In the form of the Osiris saga the dismemberment

appears, however, not merely as emasculation. More clearly recognizable is also the separation of the primal parents, the dying out of the primal being resulting in a release of the primal procreative power for a fresh world creation.   It is a very interesting point that in one of the versions a mighty tree grows out of the corpse of Osiris.   Later on we become acquainted for the first time with the potent motive of the restoration of the dismembered one, the revivification of the dead.

For example, in the Finnish epic, Kalevala, Nasshut throws the Lemminkainen into the waters of the river of the dead.   Lemminkainen was dismembered, but his mother fished out the pieces, one of which was missing, put them together and brought them to life in her womb.   According to Stucken's explanation we recognize in Nasshut a father image, in Lemminkainen a son image.   In the tradition no relationship between them is mentioned.   That is, however, a " Differentiation and attenuation of traits, which is common in every myth-maker." (S. A. M., p. 107.)

In the Edda it is recounted " that Thor fared forth with his chariot and his goats and with him the Ase, called Loki.   They came at evening to a peasant and found shelter with him.   At night Thor took his goats and slew them; thereupon they were skinned and put into a kettle.   And when they were boiled Thor sat down with his fellow travelers to supper.   Thor invited the peasant and his wife and two children to eat with him.   The peasant's son

was called Thialfi and the daughter Roskwa.   Then
Thor laid the goats' skins near the hearth and said
that the peasant and his family should throw the
bones onto the skins.   Thialfi, the peasant's son,
had the thigh bone of one goat and cut it in two
with his knife to get the marrow.   Thor stayed
there that night, and in the morning he got up before
dawn, dressed, took the hammer, Miolner, and lifted
it to consecrate the goats' skins.   Thereupon the
goats stood up; but one of them was lame in the
hind leg.   He noticed it and said that the peasant
or some of his household must have been careless
with the goats' bones, for he saw that a thigh bone
was broken."   We are especially to note here that
the hammer is a phallic symbol.

In fairy tales the dismemberments and revivifications occur frequently.   For example, in the tale of
the Juniper Tree [Machandelboom] (Grimm, K. H.
M., No. 47), a young man is beheaded, dismembered, cooked and served up to his father to be
eaten.   The father finds the dish exceptionally good.
On asking for his son he is answered that the youth
has gone to visit relatives.   The father throws all
the bones under the table.   They are collected by
the sister, wrapped in a bit of cloth and laid under
the juniper tree.   The soul of the boy soared in
the air as a bird and was afterward translated into
a living youth.   The Grimm brothers introduce as
a parallel:   " The collection of the bones occurs in
the myths of Osiris and Orpheus, and in the legend
of Adelbert; the revivification in many others, e.g.,

in the tale of Brother Lustig (K. H. M., No. 81),
of Fichter's Vogel (No. 46), in the old Danish song
of the Maribo-Spring, in the German saga of the
drowned child, etc." Moreover, W. Mannhardt
(Germ. Mythen., pp. 57–75) has collected numerous sagas and fairy tales of this kind, in which
occur the revivifications of dismembered cattle, fish,
goats, rams, birds, and men.

The gruesome meal in the story of the juniper tree
reminds us of the Tantalus story and the meal of
Thyestes. Demeter (or Thetes) ate a shoulder of
the dismembered Pelops, who was set before the
gods by his father Tantalus, and the shoulder, after
he was brought to life again, was replaced by an
ivory one. In a story from the northeastern Caucasus, a chamois similarly dismembered and brought
to life, like Thor's goats, gets an artificial shoulder
(of wood).

For the purpose of being brought to life again
the parts of the dismembered animal are regularly
put in a vessel or some container (kettle, box, cloth,
skin). In the case of the kettle, which corresponds
to the belly or uterus, they are generally cooked.
Thus in the tale of the juniper tree, the magic re-
juvenations of Medea, which — except in the version mentioning the magic potion — she practices
on Jason and Æson, and also on goats (cf. Thor
and his goats). I must quote still other pertinent
observations of Rank (p. 313 ff). The motive of
revivification, most intimately connected with dismemberment, appears not only in a secondary rôle

to compensate for the killing, but represents as well simple coming to life, i.e., birth. Rank believes that coming to life again applies originally to a dissected snake (later other animals, chiefly birds), in which we easily recognize the symbolical compensation for the phallus of the Osiris story, excised and unfit for procreation, which can be brought to life again by means of the water of life. " The idea that man himself at procreation or at birth is assembled from separate parts, has found expression not only in the typical widespread sexual theories of children, but in countless stories (e.g., Balzac's Contes drolatiques) and mythical traditions. Of special interest to us is the antique expression communicated by Mannhardt (Germ. Myth., p. 305), which says of a pregnant woman that she has a belly full of bones," which strikingly suggests the feature emphasized in all traditions that the bones of the dismembered person are thrown on a heap, or into a kettle (belly) or wrapped in a cloth. [Even the dead Jesus, who is to live again, is enveloped in a cloth. In several points he answers the requirements of the true rejuvenation myth. The point is also made that the limbs that are being put in the cloth must be intact, so that the resurrection may be properly attained (as in a bird story where the dead bird's bones must be carefully preserved). The incompleteness (stigmata) also appears after the resurrection.

John XIX, 33. " But when they came to Jesus, and saw that he was dead already, they brake not

his legs."— 40 f.  " Then took they the body of
Jesus and wound it in linen cloths with the spices.
. . . Now in the place where he was crucified there
was a garden, and in the garden a new sepulcher,
wherein was never man yet laid."

We shall mention later the significance of garden
and grave.   It supports that of the cloths.]

Rank considers that the circumstance that the dismembered person or animal resurrected generally
lacks a member, points without exception to castration.

What he has said about dismemberment we can
now sum up with reference to the lion in the parable
in the formulae: Separation of parents; the pushing
aside of the father; castration of father; taking his
place; liberation of the power of procreation; improvement.   In its bearing on the incest wish, castration is indeed the best translation of the " anatomizing " of the lion.   The dragon fighter has to
release a woman.   The idea that the mother is in
need of being released, and that it is a good deed
to free her from her oppressor, father, is according
to the insight of psychoanalysis a typical element
of those unconscious phantasies of mankind, which
are stamped deeply with the greatest significance in
the imaginative " family romance " of neurotics.
To the typical dragon fight belongs, however (according to Stucken's correct formulation), the motive of denial.   As a matter of fact the hero of our
parable is denied the prize set before him — the
admission into the college — for several of the

elders insist on the condition that the wanderer must resuscitate the lion (Sec. 7). In myths where the dragon has to fight with a number of persons this difference generally occurs: that he produces dissension among his opponents. (Jason throws a stone among the men of the dragon's teeth, they fight about the stone and lay each other low.) Dissension occurs also among the old men. They turned (Sec. 7) "fiercely on each other" if only with words.

The wanderer removes, as it were, an obstacle by the fight, tears down a wall or a restraint. This symbol occurs frequently in dreams; flying or jumping over walls has a similar meaning. The wanderer was carried as if in flight to the top of the wall. Then first returns the hesitation. The symbolism of the two paths, right and left, has already been mentioned. The man that precedes the wanderer (Sec. 7 and 8) may be quite properly taken as the father image; once, at any rate, because the wanderer finds himself on the journey to the mother (that is indeed the trend of the dream) and on this path the father is naturally the predecessor. The father is, however, the instructor, too, held up as an example and as a model for choosing the right path. The father follows the right path to the mother also; he is the lawful husband. The son can reach her only on the left path. This he takes, still for the purpose of making things better. Some one follows the wanderer on the other side (Sec. 8), whether man or woman is not known. The father image in front of the wanderer is his future for he will occupy his father's place.   The Being behind him is surely the past, the careless childhood, that has not yet learned the difference between man and woman.   It does not take the difficult right way, but quite intelligibly, the left.   The wanderer himself turns back to his childish irresponsibility; he takes the left path.   The many people that fall down may be a foil to illustrate the dangers of the path, for the purpose of deepening the impression of improvement.   Phantasies of extraordinary abilities, special powers; contrasts to the anxiety of examinations; all these in the case of the wanderer mark the change from apprehension to fulfillment. We must not fail to recognize the element of desire for honor; it will be yet described.   In view of myth motives reported by Stucken, the entire wall episode is to be conceived as a magic flight; the people that fall off are the pursuers.

At the beginning of the ninth section of the parabola, the wanderer breaks red and white roses from the rosebush and sticks them in his hat.   Red-white we already know as sexuality.   The breaking off of flowers, etc., in dreams generally signifies masturbation; common speech also knows this as " pulling off " or " jerking off."   In the symbolism of dreams and of myths the hat is usually the phallus.   This fact alone would be hardly worth mentioning, but there are also other features that have a similar significance.   The fear of impotence points to auto-erotic components in the psychosexual constitution of the wanderer (of course not clearly recognized as such), which is shown as well in the anxiety about ridicule and disgrace that awaken ambition. This is clearest in the paragraphs 6, 10, 14, of the parable. That the masturbatory symbol precedes the subsequent garden episode, can be understood if we realize that the masturbation phantasy (which has an enormous psychic importance) animates or predetermines the immediately following incest.

The wall about the garden that makes the long detour necessary (Sec. 9) is as we know the resistance. Overcoming the resistance = going round the wall, removal of the wall. Of course, after the completion of the detour there is no wall. The wall, however, signifies also the inaccessibility or virginity of the woman. The wall surrounds a garden. The garden is, however (apart from the paradise symbolism derived from it), one of the oldest and most indubitable symbols for the female body.

> " Maiden shall I go with you
> In your rose garden,
> There where the roses stand
> The delicate and the tender;
> And a tree nearby
> That moves its leaves,
> And a cool spring
> That lies just under it."

Without much change the same symbolism is found in stately form in the Melker Marienlied of the 12th century. (See Jung, Jb. ps. F., IV, S. 398 ff.)

Sainted Mary
Closed gate
Opened by God's word —
Sealed fountain,
Barred garden,
Gate of Paradise.

Note also the garden, roses and fountains in the Song of Solomon.

The wanderer wishes to possess his mother as an unravished bride.    Also a feature familiar to psychoanalysis.    The generally accompanying antithesis is the phantasy that the mother is a loose woman = attainable, sexually alluring woman.    Perhaps this idea will also be found in the parable.

The young people of both sexes, separated by a wall, do not come together because they are afraid of the distant detour to the door.    This can, with a little courage, be translated: The auto-erotic satisfaction is easier.

[C. G. Jung writes (Jb. ps. F., IV, p. 213 ff): " Masturbation is of inestimable importance psychologically.    One is guarded from fate, since there is no sexual need of submitting to any one, life and its difficulties.    With masturbation one has in his hands the great magic.    He needs only to imagine and in addition to masturbate and he possesses all the pleasures of the world and is under no compulsion to conquer the world of his desire through hard work and struggle with reality.    Aladdin rubs his lamp and the slaves come at his bidding; this story expresses the great psychologic gain in local sexual sat-

isfaction through facile regression." Jung applies
to masturbation the motive of the dearly won prize
and that of the stealing of fire.   He even appears to
derive in some way the use of fire from masturbation.   In this at any rate I cannot follow him.]

On his detour the wanderer (who desires to reach
the portal of woman) meets people who are alone in
the rooms and carry on dirty work.   Dirt and masturbation are wont to be closely associated psychically.   The dirty work is " only appearance and individual fantasy," and " has no foundation in Nature."   The wanderer knows that " such practices
vanish like smoke."   He has done it himself before
and now he will have nothing more to do with it.
He aspires to a woman, that the work done alone
leads to nothing is connected with the fact that the
work of two is useful.   But " dirty work " is also to
be understood as sensual enjoyment without love.

In paragraph 10 we again meet the already mentioned symbolism of the walled garden.   The wanderer is the only one that can secure admission to
the maiden.   After a fear of impotence (anxiety
about disgrace) he goes resolutely to the door and
opens it with his Diederich, which sticks into a narrow, hardly visible opening (deflowering).   He
" knows the situation of the place," although he has
never been there before.   I mean that once, before
he was himself, he was there in the body of his
mother.   What follows suggests a birth fantasy as
these occur in dreams of being born.   The wanderer
now actually takes part in being born in reverse di-

rection. I append several dreams about being born.

"I find myself on a very narrow stairway, leading down in turns; a winding stairway. I turn and push through laboriously. Finally I find a little door that leads me into the open, on a green meadow, where I rest in soft luxuriant bushes. The warm sunshine was very pleasant."

F. S. dreams: "In the morning I went to work with my brother (as we went the same road) in the Customs House Street. Before the customs house I saw the head postillion standing. From it the way led to a street between two wooden walls; the way appeared very long and seemed to get narrower toward the end and indeed so close that I was afraid that we would not get through. I went out first, my brother behind me; I was glad when I got out of the passage and woke with a beating heart." Addenda. "The way was very dark, more like a mine. We couldn't see, except in the distance the end, like a light in a mine shaft. I closed my eyes." Stekel notes on the dreams of F. S.: "The dream is a typical birth dream. The head postillion is the father. The dreamer wants to reverse the birth relations of his brother who is ten years older than he. ' I went out first, my brother after me.' "

Another beautiful example in Stekel: "Inter faeces et urinas nascimur." [We are born between faeces and urine] says St. Augustine. Mr. F. Z. S. contributes an account of his birth which strongly reminds us of the sewer-theory.

"I went into the office and had to pass a long,

narrow, uneven alley. The alley was like a long court between two houses and I had the indefinite feeling that there was no thoroughfare. Yet I hurried through. Suddenly a window over me opened and some one, I believe a female, spilled the wet contents of a vessel on me. My hat was quite wet and afterwards I looked at it closer and still noticed the traces of a dirty gray liquid. Nevertheless, I went on without stopping, and quickened my steps. At the end of the alley I had to go through the house that was connected by the alley with the other. Here I found an establishment (inn?) that I passed. In this establishment were people (porters, servants, etc.) engaged in moving heavy pieces of furniture, etc., as if these were being moved out or rearranged. I had to be careful and force my way through. Finally I came to the open on a street and looked for an electric. Then I saw on a path that went off at an angle, a man whom I took for an innkeeper who was occupied in measuring or fastening a hedge or a trellis. I did not know exactly what he was doing. He was counting or muttering and was so drunk that he staggered."

Stekel: " In this dream are united birth and effects of the forbidden or unpermissible. The dreamer goes back over the path — evidently as an adult. The experiences represent an accusation against the mother. This accusation was not without reason. Mr. F. Z. S. had a joyless childhood. His mother was a heavy drinker. He witnessed her

coitus with strangers. (Packing up = coitus.)
The packers and porters are the strange men who
visited his inn (his mother was also his nurse) in
order to store heavy objects, etc. Finally he was
obstructed in his birth, for a man is occupied in
measuring. The father was a surveyor (the innkeeper). In the dream, furthermore, he was measuring a trellis-fence. Both trellis and hedge are
typical dream symbols of obstacles to copulation."

Comparing Sections 10 and 11 of the parable
carefully with the contents of this dream, we find
astonishing correspondences. Notice the details,
e.g., the rosebushes, the sun, the rain, the hedge.
On the " well builded house " of Section 10, I shall
only remark that Scherner has noticed in this connection that the human body represents a building.
" Well built house " signifies " beautiful body."

If we remember that the wanderer reverses the
way of birth, we shall not be surprised that he finds
a smaller garden in the larger. That is probably
the uterus. The wanderer attains the most intimate
union with his ideal, the mother, in imagining himself in her body. This phantasy is continued still
less ambiguously,— but I do not wish to anticipate.
Be it only said: He possesses his mother as a spouse
and as a child; it is as if in the desire to do everything better than his father he desires to beget himself anew. We already know the mythological motives of new creation, that should follow the forcible
separation of the parents and that we have not yet

noticed in the parable.   Shall the better world still be created, the dismembered paternal power be renewed, the lion again be brought to life?

The rectangular place in the garden suggests a grave.   A wall in a dream means, among other things, a cemetery wall and the garden, a cemetery. And widely as these ideas may be contrasted with the lifegiving mother's womb, they yet belong psychologically in very close connection with her.   And perhaps not only psychologically.

Stekel tells a dream of Mrs. Delta in which occurs " an open square space, a garden or court.   In the corner stood a tree, that slowly sinks before our eyes as if it were sinking in water.   As the tree and the court also made swinging motions, I cleverly remarked, ' Thus we see how the change in the earth's surface takes place.' "   The topmost psychic stratum of the dream reveals itself as an earthquake reminiscence.   " Earth " leads to the idea of " Mother Earth."   The tree sinking into it, is the tree of life, the phallus.   The rectangular space is the bedroom, the marriage bed.   The swinging motions characterize the whole picture still better. The earthquake, however, contains, as is found in the analysis, death thoughts also.   The rectangular space becomes a grave.   Even the water of the dream deserves notice.   " Babies come out of the water," says an infantile theory of procreation. We learn later that the fœtus floats in amniotic liquor.   This " water " lies naturally in " Mother Earth."   In contrast we have the water of the dead (river of the dead, islands of the dead, etc.). Both waters are analogous in the natural symbolism. It is the mythical abode of the people not yet, or no longer, to be found in the world.

As water will appear again at important points in the parable, I will dwell a little longer on that topic.

Little children come out of Holla's fountain, there are in German districts a number of Holla wells or Holla springs (Holla brunn?) with appropriate legends. Women, we are told, who step into those springs become prolific. Mullenhof tells of an old stone fountain in Flensburg, which is called the Grönnerkeel. Its clear, copious water falls out of four cocks into a wide basin and supplies a great part of the city. The Flensburgers hold this fountain in great honor, for in this city it is not the stork which brings babies, but they are fished out of this fountain. While fishing the women catch cold and therefore have to stay in bed. Bechstein (Fränk. Sagensch.) mentions a Little Linden Spring on a road in Schweinfurt near Königshofen. The nurses dip the babies out of it with silver pails, and it flows not with water but with milk. If the little ones come to this baby fountain they look through the holes of the millstone (specially mentioned on account of what follows) at its still water, that mirrors their features, and think they have seen a little brother or sister that looks just like them. (Nork. Myth d. Volkss., p. 501.)

From the lower Austrian peasantry Rank takes the following (Wurth. Zf. d. Mythol., IV, 140):

" Far, far off in the sea there stands a tree near which the babies grow.   They hang by a string on the tree and when the baby is ripe the string breaks and the baby floats off.   But in order that it should not drown, it is in a box and in this it floats away to the sea until it comes into a brook.   Now our Lord God makes ill a woman for whom he intended the baby.   So a doctor is summoned.   Our Lord God has already suggested to him that the sick woman will have a baby.   So he goes out to the brook and watches for a long time until finally the box with the baby comes floating in, and he takes it up and brings it to the sick woman.   And this is the way all people get their babies."

I call attention briefly also to the legend of the fountain of youth, to the mythical and naturalistic ideas of water as the first element and source of all life, and to the drink of the gods (soma, etc.) Compare also the fountain in the verses previously quoted.

The bridal pair in the parable (Sec. 11) walk through the garden and the bride says they intend in their chamber to " enjoy the pleasures of love." They have picked many fragrant roses.   Bear in mind the picking of strawberries in Mr. T.'s dream. The garden becomes a bridal chamber.   The rain mentioned somewhat earlier, is a fructifying rain; it is the water of life that drops down upon Mother Earth.   It is identical with the sinking trees of Mrs. Delta's dream, with the power of creation developed by the wanderer, with the mythical drink of the gods, ambrosia, soma.  We shall now see the wanderer ascend or descend to the source of this water of life.  To gain the water of life it is generally in myths necessary to go down into the underworld (Ishtar's Hell Journey), into the belly of a monster or the like.  Remember, too, that the wanderer puts himself back in his mother's womb.  There is indeed the origin of his life.  The process is still more significantly worked out in the parable.

The wanderer (Section 11, after the garden episode) comes to the mill.  The water of the mill stream also plays a significant part in the sequel. The reader will surely have already recognized what kind of a mill, what kind of water is meant.  I will rest satisfied with the mere mention of several facts from folklore and dream-life.

Nork (Myth. d. Volkss., p. 301 f.) writes " that Fenja is of the female sex in the myth (Horwendil) which we must infer from her occupation, for in antiquity when only hand mills were as yet in use, women exclusively did this work.  In symbolic language, however, the mill signifies the female organ (μυλλός from which comes *mulier*) and as the man is the miller, the satirist Petronius uses *molere mulierem* = (grind a woman) for coitus, and Theocritus (Idyll, IV, 48) uses μύλλω (I grind) in the same sense.  Samson, robbed of his strength by the harlot, has to grind in the mill (Judges XVI, 21) on which the Talmud (Sota fol. 10) comments as follows:  By the grinding is always meant the sin of fornication (Beischlaf).  Therefore all the mills in Rome stand still at the festival of the chaste Vesta. Like Apollo, Zeus, too, was a miller (μυλεύς, Lykophron, 435), but hardly a miller by profession, but only in so far as he presides over the creative lifegiving principle of the propagation of creatures. It is now demonstrated that every man is a miller and every woman a mill, from which alone it may be conceived that every marriage is a milling (jede Vermählung eine Vermehlung), etc. Milling (vermehlung) is connected with the Roman confarreatio (a form of marriage); at engagements the Romans used to mingle two piles of meal. In the same author (p. 303 and p. 530): Fengo is therefore the personification of grinding, the mill (Grotti) is his wife Gerutha, the mother of Amleth or Hamlet. Grotti means both woman and mill. Greeth is only a paraphrase of woman. He continues, " Duke Otto, Ludwig of Bavaria's youngest son, wasted his substance with a beautiful miller's daughter named Margaret, and lived in Castle Wolfstein. . . . This mill is still called the Gretel mill and Prince Otto the Finner (Grimm, D. S., No. 496). Finner means, like Fengo, the miller [Fenja — old Norman? = the milleress], for the marriage is a milling [Vermählung ist eine Vermehlung], the child is the ground grain, the meal.

The same writer (Sitt. u. Gebr., p. 162): " In concept the seed corn has the same value as the spermatozoon. The man is the miller, the woman the mill."

In Dulaure-Krauss-Rieskel (Zeugung i. Glaub

usw. d. Völk., p. 100 ff.) I find the following charm
from the writings of Burkhard, Bishop of Worms:
" Have you not done what some women are accustomed to do? (They strip themselves of clothes,
they anoint their naked bodies with honey, spread
a cloth on the ground, on which they scatter grain,
roll about in it again and again, then collect carefully all the grains, which have stuck on their bodies,
and grind them on the mill stone which they turn in
a contrary direction.   When the corn is ground into
meal, they bake a loaf of it, and give it to their
husbands to eat, so that they become sick and die.
When you have done this you will atone for it forty
days on bread and water."

Killing is the opposite of procreating, therefore
the mill is here turned in reverse direction.

Etymologically it is here to be noted that the verb
mahlen (grind), iterative form of mähen (mow),
originally had a meaning of moving oneself forwards and backwards. Mulieren or mahlen
(grind), *molere,* μυλλειν for coire (cf. Anthropo-
phyteia, VIII, p. 14).

There are numerous stories where the mill appears as the place of love adventures.   The " old
woman's mill " also is familiar; old women go in
and come out young.   They are, as it were, ground
over in the magic mill.   The idea of recreation in
the womb lies at the bottom of it, just as in the
vulgar expression, " Lassen sie sich umvögeln."

In a legend of the Transylvanian Gypsies, " there
came again an old woman to the king and said:

'Give me a piece of bread, for seven times already
has the sun gone down without my having eaten
anything!' The King replied: 'Good, but I will
first have meal ground for you,' and he called his
servants and had the old woman sawn into pieces.
[Then the old woman's sawn up body changed into a
good Urme (fairy) and she soared up into the
air. . . ." (H. V. Wlislocki, Märchen u. Sagen
d. transylv. Zigeuner.)

A dream: " I came into a mill and into ever narrower apartments till finally I had no more space.
I was terribly anxious and awoke in terror." A
birth phantasy or uterus phantasy.

Another dream (Stekel, Spr. d. Tr., p. 398 f.):
" I came through a crack between two boards out
of the 'wheel room.' The walls dripped with
water. Right before me is a brook in which stands
a rickety, black piano. I use it to cross over the
brook, as I am running away. Behind me is a
crowd of men. In front of them all is my uncle.
He encourages them to pursue me and roars and
yells. The men have mountain sticks, which they
occasionally throw at me. The road goes through
the verdure up and down hill. The path is strewn
with coal cinders and therefore black. I had to
struggle terribly to gain any ground. I had to push
myself to move forwards. Often I seemed as
though grown to the ground and the pursuers came
ever nearer. Suddenly I am able to fly. I fly into
a mill through the window. In it is a space with
board walls; on the opposite wall is a large crank.

I sit on the handle, hold on to it with my hands, and fly up. When the crank is up I press it down with my weight, and so set the mill in motion. While so engaged I am quite naked. I look like a cupid. I beg the miller to let me stay here, promising to move the mill in the manner indicated. He sent me away and I have to fly out of another window again. Outside there comes along the 'Flying Post.' I place myself in front near the driver. I was soon requested to pay, but I have only three heller with me. So the conductor says to me, 'Well, if you can't pay, then you must put up with our sweaty feet.' Now, as if by command, all the passengers in the coach drew off a shoe and each held a sweaty foot in front of my nose."

This dream, too (beside other things), contains a womb phantasy, wheel room, mill, space with wet walls — the womb. The dreamer is followed by a crowd; just as our wanderer is met by a crowd; the elders. This dream, which will still further occupy our attention, I shall call the " Flying Post."

Let us return to the parable. The mill of Section 11 is the womb. The wanderer strives for the most intimate union with his mother; his striving, to do better than his father culminates in his procreating himself, the son, again and better.

He will quite fill up his mother — be the father in full. Of course the phantasy does not progress without psychic obstructions. The anxious passage over the narrow plank manifests it.

We have here the familiar obstructions to move-

ment and in a form indeed that recalls the dangerous path on the wall.  The passage over the water
is also a death symbol.  We have not only the
anxiety about death caused by the moral conflict,
but we have also to remember that the passage into
the uterus is a passage to the beyond.  The water
is the Water of Death (stygian waters) and of Life.
In narrower sense it is also seminal fluid and the
amniotic liquor.  It is overdetermined as indeed all
symbols are.  The water bears the death color =
black.  In the Flying Post dream a black road appears.  The dreamer has conflicts like those of the
wanderer.

The old miller who will give no information is
the father.  Of course he will not let him have his
mother, and he gives him no information as to the
mill work or the procreative activity.  The wheels
are, on the one hand, the organs that grind out the
child (producing the child like meal), and on the
other hand they are the ten commandments whose
mundane administration is the duty of the father,
by means of strict education and punishment.  In
passing over the plank, the wanderer places himself
above the ten commandments and above the privi-
leges of the father.  The wanderer always extri-
cates himself successfully from the difficulties.  The
anxiety is soon done away with, and the fulfillment
phase supervenes.  It is only a faint echo of the paternal commandments when the elders (immediately
after the episode, Section 11) hold out before him
the letter from the faculty.  At bottom, in retaining

their authority, they do indeed go against his own wishes (also a typical artifice of the dream technique).

I have already discussed the letter episode sufficiently (as also Sections 12 and 13), so I need say no more about the incest wish there expressed.

The bridal pair were put (Section 14) into their crystal prison. We have been looking for the reassembling of the dismembered; it takes place before our eyes, the white and red parts, bones and blood, are indeed bridegroom and bride. The prison is the skin or the receptacle in which, as in myths, the revivification takes place. Not in the sense of the revivification of the annihilated father, but a recreation (improvement) that the son accomplishes, although the creative force as such remains the same. The son "marries and mills" (vermählt und vermehlt) with his mother, for the crystal container is again the same as the mill; the uterus. Even the amniotic fluid and the nutritive liquid for the fœtus are present, and the wanderer remakes himself into a splendid king. He can really do it better than his father. The dream carries the wish fulfillment to the uttermost limits.

Let us examine the process somewhat more in detail. The wanderer, by virtue of a dissociation, has a twofold existence, once as a youth in the inside of the glass sphere, and once outside in his former guise. Outside and inside he is united with his mother as husband and as developing child. He there embraces his " sister " (image of his mother

renewed with him as it were) as Osiris does his
sister Isis.   And in addition to this the infantile
sexual components of exhibitionism find satisfaction,
for whose gratification the covering of the procreation mystery is made of glass.   The sexual influence
of the wanderer on the kettle (uterus) is symbolically indicated by the fire task allotted to him.   The
fire is one of the most frequent love symbols in
dreams.   Language also is wont to speak of the fire
of love, of the consuming flames of passion, of ardent desires, etc.   Customs, in particular marriage
customs, show a similar symbolism.   That the wanderer is charged with a duty, and explicitly com-
manded to do what he is willing to do without orders, is again the already mentioned cunning device
of the dream technique to bring together the incom-
patible.   It seems almost humorous when the prison
is locked with the seal of the right honorable faculty;
I recall to you the expression " sealing " (petsch-
ieren) ; the sealing is an applying of the father's
penis.   In the place of father we find, of course, the
officiating wanderer.   The sealing means, however,
the shutting up of the seed of life that is placed in
the mother.   It is also said that the pair, after the
confinement in the prison, can be given no more
nourishment; and that the food with which they are
provided comes exclusively from the water of the
mill.   That refers to the intrauterine nourishment,
to which nothing, of course, can be supplied but the
water of the mill so familiar to us.

The precious vessel that the wanderer guards is

surrounded by strong walls; it is inaccessible to the others; he alone may approach with his fire. It is winter. That is not merely a rationalizing (pretext of commonplace argument) of the firing, but a token of death entering into the uterus. The amorous pair in the prison dissolve and perish, even rot (Section 15). I must mention incidentally, for the understanding of this version, that at the time of the writing of the parable the process of impregnation was associated with the idea of the " decaying " or " rotting " of the semen. The womb is compared to the earth in which the kernel of grain " decays."

The decaying which precedes the arising of the new being is connected with a great inundation. Mythically, a deluge is actually accustomed to introduce a (improved) creation. A proper myth can hardly dispense with the idea of a primal flood. I would, in passing, note that the present phase of the parable corresponds mythologically to the motive of being swallowed, the later release from the prison is the spitting forth (from the jaws of the monster), the return from the underworld. The dismemberment motive of the cosmogonies is usually associated with a deluge motive. In the description of the flood in the parable there are, moreover, included some traits of the biblical narrative, e.g., the forty days and the rainbow. This, be it remarked in passing, had appeared before; it is a sign of a covenant. It binds heaven and earth, man and woman. The flood originates in the falling of tears; it arises also from the body of the woman; it refers to the well known highly significant water. Stekel has arranged for dreams the so-called symbolic parallels, according to which all secretions and excretions may symbolically represent each other. On the presupposition that marks of similarity are not conceived in a strict sense, the following comparisons may be drawn:  Mucus = blood = pus = urine = stools = semen = milk = sweat = tears = spirit = air = [breath = flatus] = speech = money = poison.   That in this comparison both souls and tears appear is particularly interesting; the living or procreating principle appears as soul in the form of clouds.   These are formed from water, the Water of Life.   The dew that comes from it impregnates the earth.

As we have now reached the excreta, I should like to remind the reader of the foul and stinking bodies that in the parable lie in liquid (Section 15) on which falls a warmer rain.   The parable psychoanalytically regarded, is the result of a regression leading us into infantile thinking and feeling; we have seen it clearly enough in the comparison with the myths.   And here it is to be noticed how great an interest children take in the process of defecation. I should not have considered this worthy of notice, did not the hermetic symbolism, as we shall see later, actually use in parallel cases the expressions " fimus," " urina puerorum," etc., in quite an unmistakable manner.   In any case it is worth remembering that out of dung and urine, things that decompose malodorously and repulsively, fresh life arises.   This agrees with the infantile theory of procreation, that babies are brought forth as the residue of assimilation; we are to observe, however, still other interrelations that will be encountered later. A series of mythological parallels may be cited. I shall rest satisfied with referring to the droll story, " Der Dumme Hans." Stupid Jack loads manure (fæces, sewage) into a cart and goes with it to a manor; there he tells them he comes from the *Moorish* land (from the country of the blacks) and carries in his barrel the *Water of Life*. When any one opens the barrel without permission, Stupid Jack represented himself as having turned the water of life into sewage. He repeated the little trick with his *dead* grandmother whom he sewed up in black cloths and gave out as a wonderfully beautiful princess who was lying in a hundred years' sleep. Again, as he expected, the covering was raised by an unbidden hand and John lamented, that, on account of the interference, instead of the princess, whom he wanted to take to the King, a disgusting corpse had been magically substituted. He succeeded in being recompensed with a *good deal* of money. [Jos. Haltrich, Deut. Volksmed. d. Siebenbürg, II, p. 224.]

Inasmuch as the wanderer of our parable finds himself not outside but inside of the receptacle, he is as if in a bath. I note incidentally that writings analogous to the parable expressly mention a bath in a similar place, as the parable also does (Sec. 15). In dreams the image of bathing frequently appears to occur as a womb or birth phantasy.

At the end of the 14th section, as the inmates of the prison die, his certain ruin stands before the wanderer's eyes — again a faint echo of his relation to the bridegroom.

We have already for a long time thoroughly familiarized ourselves with the thought that in the crystal prison the revivification of the dismembered comes to pass. Whoever has the slightest doubt of it, can find it most beautifully shown in the beginning of Section 15. The author of the parable even mentions Medea and Æson. I need add nothing more concerning the talents of the Colchian sorceress in the art of dissection and rejuvenation.

In Section 18, " the sun shines very bright, and the day becomes warmer than before and the dog days are at hand." Soon after (Sec. 19) the king is released from prison. It was before the winter (Sec. 14), but after that season, when the sun " shines very warm " (Sec. 11), consequently well advanced into autumn. Let us choose for the purpose a middle point between the departing summer and the approaching winter, about the end of October, and bear in mind that the dog-days come in August, so that at the end of July they are in waiting, then we find for the time spent in the receptacle nine months — the time of human gestation.

The newborn (Sec. 20) is naturally — thirsty. What shall he be fed with if not with the water from the mill? And the water makes him grow and thrive.

Two royal personages stand before us in splendor

and magnificence.    The wanderer has created for himself new parents (the father-king is, of course, also himself) corresponding to the family romance of neurotics, a phantasy romance, that like a ghost stalks even in the mental life of healthy persons. It is a wish phantasy that culminates in its most outspoken form in the conviction that one really springs from royal or distinguished stock and has merely been found by the actual parents who do not fit. They conceal his true origin.    The day will come, however, when he will be restored to the noble station which belongs to him by right.    Here belong in brief, those unrestrained wish phantasies which, no matter in what concrete form, diversify the naïvely outlined content.    They arise from dissatisfaction with surroundings and afford the most agreeable contrasts to straitened circumstances or poverty. In the parable especially, the King (in his father character) is attractively portrayed.

At first the " lofty appearance " (Sec. 19) of the severe father amazes the wanderer, then it turns out, however, that the king (ideal father) is friendly, gracious and meek, and we are assured that " nothing graces exalted persons as much as these virtues." And then he leads the wanderer into his kingdom and allows him to enjoy all the merely earthly treasures.    There takes place, so to speak, a universal gratification of all wishes.

Mythologically we should expect that the hero thrown up from the underworld, should have brought with him the drink of knowledge.    This is

actually the case, as he has indeed gained the thing
whose constitution is metaphorically worked out in
the whole story, that is, the philosopher's stone.
The wanderer is a true soma robber.

Let us hark back to the next to last section.
Here, near the end of the dream, the King becomes
sleepy.   The real sleeper already feels the approaching awakening and would like to sleep longer (to
phantasy).   But he pretends that the king is sleepy,
thus throwing the burden from his own shoulders.
And to this experience is soon attached a symbol of
waking: the wanderer, the dreamer of the parable,
is taken to another land, indeed into a bright land.
He wakes from his dreams with a pious echo of his
wish fulfillment on his lips . . . " to which end
help us, the Father, Son, and Holy Ghost, Amen."
It is quite prosaic to conclude this melodious finale
by means of the formula " threshold symbolism."

To sum up in a few words what the parable contains from the psychoanalytic point of view, and to
do this without becoming too general in suggesting
as its results the universal fulfillment of all wishes,
I should put it thus: the wanderer in his phantasy
removes and improves the father, wins the mother,
procreates himself with her, enjoys her love even in
the womb and satisfies besides his infantile curiosity
while observing procreative process from the outside.   He becomes King and attains power and magnificence, even superhuman abilities.

Possibly one may be surprised at so much absurdity.   One should reflect, however, that those

unconscious titanic powers of imagination that, from the innermost recesses of the soul set in motion the blindly creating dream phantasy, can only wish and do nothing but wish.   They do not bother about whether the wishes are sensible or absurd.   Critical power does not belong to them.   This is the task of logical thinking as we consciously exercise it, inasmuch as we observe the wishes rising from the darkness and compare and weigh them according to teleological standards.   The unconsciously impelling affective life, however, desires blindly, and troubles itself about nothing else.

## Section II

## ALCHEMY

THE tradition of craftsmanship in metallurgy, an art that was practiced from the earliest times, was during the speculative period of human culture, saturated with philosophy. Especially was this the case in Egypt, where metallurgy, as the source of royal riches and especially the methods of gold mining and extraction, were guarded as a royal secret. In the Hellenistic period the art of metal working, knowledge of which has spread abroad and in which the interest had been raised to almost scientific character, was penetrated by the philosophical theories of the Greeks: the element and atom ideas of the nature-philosophers and of Plato and of Aristotle, and the religious views of the neoplatonists. The magic of the orient was amalgamated with it, Christian elements were added — in brief, the content of the chemistry of that time, which mainly had metallurgy as its starting point, took a vital part in the hybrid thought of syncretism in the first centuries after Christ.

As the chemical science (in alchemy, alkimia, al is the Arabic article prefixed to the Greek χημεία) has come to us from the Arabs (Syrians, Jews, etc.) it was long believed that it had an Arabian origin.

Yet it was found later that the Arabs, while they added much of their own to it, still were but the preservers of Greek-Hellenistic knowledge and we are convinced that the alchemists were right when they indicated in their traditions the legendary Egyptian Hermes as their ancestor. This legendary personage is really the Egyptian god Thoth, who was identified with Hermes in the time of the Ptolemies. He was honored as the Lord of the highest wisdom and it was a favorite practice to assign to him the authorship of philosophical and especially of theological works. Hermes' congregations were formed to practice the cult, and they had their special Hermes literature.[1] In later times the divine, regal, Hermes figure was reduced to that of a magician. When I speak, in what follows, of the hermetic writings I mean (following the above mentioned traditions) the alchemic writings, with, however, a qualification which will be mentioned later.

The idea of the production of gold was so dominant in alchemy that it was actually spoken of as the gold maker's art. It meant the ability to make gold out of baser material, particularly out of other metals. The belief in it and in the transmutability of matter was by no means absurd, but rather it must be counted as a phase in the development of human thought. As yet unacquainted with the modern doctrine of unchangeable elements they could draw no

[1] Information on this point will be found in Reitzenstein's "Poimandres."

other conclusion from the changes in matter which they daily witnessed. If they prepared gold from ores or alloys, they thought they had " made " it. By analogy with color changes (which they produced in fabrics, glass, etc.) they could suppose that they had colored (tinctured) the baser metals into gold.

Under philosophical influences the doctrine arose that metals, like human beings, had body and soul, the soul being regarded as a finer form of corporeality. They said that the soul or primitive stuff (prima materia) was common to all metals, and in order to transmute one metal into another they had to produce a tincture of its soul. In Egypt lead, under the name Osiris, was thought to be the primitive base of metals; later when the still more plastic quicksilver (mercury) was discovered, they regarded this as the soul of metals. They thought they had to fix this volatile soul by some medium in order to get a precious metal, silver, gold.

That problematic medium, which was to serve to tincture or transmute the baser metal or its mercury to silver or gold, was called the Philosopher's stone. It had the power to make the sick (base) metal well (precious). Here came in the idea of a universal medicine. Alchemy desired indeed to produce in the Philosopher's Stone a panacea that should free mankind of all sufferings and make men young.

It will not be superfluous to mention here, that the so-called materials, substances, concepts, are found employed in the treatises of the alchemists in

a more comprehensive sense, we can even say with
more lofty implications, the more the author in question leans to philosophical speculation. The authors who indulged the loftiest flights were indeed
most treasured by the alchemists and prized as the
greatest masters. With them the concept mercury,
as element concept, is actually separated from that
of common quicksilver. On this level of speculation, quicksilver (Hg.) is no longer considered as a
primal element, but as a suprasensible principle to
which only the name of quicksilver, mercury, is
loaned. It is emphasized that the *mercurius philosophorum* may not be substituted for common quicksilver. Similar transmutations are effected by the
concept of a primal element specially separated from
mercury. Prima materia is the cause of all objects.
Also the material from which the philosopher's
stone is produced is in later times called the prima
materia, accordingly in a certain sense, the raw material (materia cruda) for its production. But I
anticipate; this belongs properly to the occidental
flourishing period of the alchemy of scholasticism.

A very significant and ancient idea in alchemy is
that of sprouting and procreation. Metals grow
like plants, and reproduce like animals. We are assured by the adepts (those who had found it, viz.,
the panacea) in the Greek-Egyptian period and also
later, that gold begets gold as the corn does corn,
and man, man. The practice connected with this
idea consists in putting some gold in the mixture that
is to be transmuted. The gold dissolves like a seed

in it and is to produce the fruit, gold. The gold ingredient was also conceived as a ferment, which permeates the whole mixture like a leaven, and, as it were, made it ferment into gold. Furthermore, the tincturing matter was conceived as male and the matter to be colored as female. Keeping in view the symbol of the corn and seed, we see that the matter into which the seed was put becomes earth and mother, in which it will germinate in order to come to fruition.

In this connection belongs also the ancient alchemic symbol of the philosopher's egg. This symbol is compared to the " Egyptian stone," and the dragon, which bites its tail; consequently the procreation symbol is compared to an eternity or cycle symbol. The " Egyptian stone " is, however, the philosopher's stone or, by metonomy, the great work (magnum opus) of its manufacture. The egg is the World Egg that recurs in so many world cosmogonies. The grand mastery refers usually and mainly to thoughts of world creation. The egg-shaped receptacle in which the master work was to be accomplished was also known as the " philosophical egg " in which the great masterpiece is produced. This vessel was sealed with the magic seal of Hermes; therefore hermetically sealed.

A wider theoretical conception, originating with the Arabs, is the doctrine of the two principles. They were retained in the subsequent developments and further expanded. Ibn Sina [Avicenna, 980– ca. 1037] taught that every metal consisted of mercury and sulphur. Naturally they do not refer to the ordinary quicksilver and ordinary sulphur.

From the Arabs alchemy came to the occident and spread extraordinarily. Among prominent authors the following may be selected: Roger Bacon, Albertus Magnus, Vincent of Beauvais, Arnold of Villanova, Thomas Aquinas, Raymond Lully, etc.

The amount of material that could be adduced is enormous. It is not necessary, however, to consider it. What I have stated about the beginnings of alchemy is sufficient in amount to enable the reader to understand the following exposition of the alchemic content of the parable. And what I must supply in addition to the alchemic theories of the time of their prevalence in the west, the reader will learn incidentally from the following analysis.

In concluding this preliminary view I must still mention one novelty that Paracelsus (1493–1541) introduced into the theory. Ibn Sina had taught that two principles entered into the constitution of metals. Mercury is the bearer of the metallic property and sulphur has the nature of the combustible and is the cause of the transmutation of metals in fire. The doctrine of the two principles leads to the theory that for the production of gold it was necessary to get from metals the purest possible sulphur and mercury, in order to produce gold by the union of both. Paracelsus now adds to the two principles a third, salt, as the element of fixedness or palpability, as he terms it. According to my notion, Paracelsus has not introduced an essential innovation, but only used in a new systematic terminology what others said before him, even if they did not follow it out so consistently. The principles mercury, sulphur and salt — their symbols are ☿, ♁ and ⊖— were among the followers of the alchemists very widely used in their technical language. They were frequently also called spirit, soul and body. They were taken in threes but also as before in twos, according to the exigencies of the symbolism.

The alchemists' usual coupling of the planets with metals is probably due to the Babylonians. I reproduce these correspondences here in the form they generally had in alchemy. I must beg the reader to impress them upon his memory, as alchemy generally speaks of the metals by their planetary names. According to the ancient view (even if not the most ancient) there are seven planets (among which was the sun) and seven metals.

| Planet. | Symbol. | Metal. |
|---------|---------|--------|
| Saturn. | ♄ | Lead. |
| Jupiter. | ♃ | Tin. |
| Mars. | ♂ | Iron. |
| Sun. | ☉ | Gold. |
| Venus. | ♀ | Copper. |
| Mercury. | ☿ | Quicksilver. |
| Moon. | ☽ | Silver. |

Relative to the technical language, which I must use in the following discussion also, I have to make a remark of general application that should be carefully remembered. It is a peculiarity of the alchemistic authors to use interchangeably fifty or more names for a thing and on the other hand to give one and the same name many meanings. This custom was originally caused partly by the uncertainty of the concepts, which has been mentioned above. But this uncertainty does not explain why, in spite of increase of knowledge, the practice was continued and purposely developed. We shall speak later of the causes that were active there. Let it first be understood merely that it was the case and later be it explained how it comes about that we can find our way in the hermetic writings in spite of the strange freedom of terminology that confuses terms purposely and constantly. Apart from a certain practice in the figurative language of the alchemists, it is necessary, so to speak, to think independently of the words used and regard them only in their context. For example, when it is written that a body is to be washed with water, another time with soap, and a third time with mercury, it is not water and soap and mercury that is the main point, but the relation of all to each other, that is the washing, and on closer inspection of the connection it can be deduced that all three times the same cleansing medium is meant, only described three times with different names.

The alchemistic interpretation of our parable is a development of what its author tried to teach by it. We do not need to show that he pursues an hermetic aim, for he says so himself, and so do the circumstances, i.e., the book, in which the parable is found. In this respect we shall fare better in the alchemistic exposition than in the psychoanalytic, where we were aiming at the unconscious. Now we have the conscious aim before us and we advance with the author, while before we worked as it were against his understanding, and deduced from the product of his mind things that his conscious personality would hardly admit, if we had him living before us; in which case we should be instructing him and informing him of the interpretation afforded by psychoanalysis.

In one respect we are therefore better off, but in another we are much worse off. For the matter in which we previously worked, the unconscious, remains approximately the same throughout great periods; the unconscious of the wanderer is in its fundamentals not very different from that of a man of to-day or from that of Zosimos. [Zosimos is one of the oldest alchemistic writers of whom we have any definite knowledge — about the 4th century.] It is the soul of the race that speaks, its " humanity." Much more swiftly, on the contrary, does objective knowledge change in the course of time and the forms also in which this knowledge is expressed. From this point of view the conscious is more difficult of access than the unconscious. And now we have to face a system so very far removed from our way of thinking as the alchemistic.

Fortunately I need not regard it as my duty to explain the parable so completely in the alchemistic sense that any one could work according to it in a chemical laboratory. It is much more suitable to our purpose if I show in general outline only how we must arrange the leading forms and processes of the parable to accord with the mode of thinking peculiar to alchemy. If I should succeed in doing so clearly, we should already have passed a difficult stage. Then for the first time I might venture further — to the special object of this research. But patience! We have not yet gone so far.

First of all it will be necessary for me to draw in a few lines a sketch of how, in the most flourishing period of alchemy, the accomplishment of the Great Work was usually described. In spite of the diversity of the representations we find certain fundamental principles which are in general firmly established. I will indicate a few points of this iron-clad order in the alchemic doctrine.

There is, in the first place, the central idea of the interaction or the coöperation of two things that are generally called man and woman, red and white, sun and moon, sulphur and mercury. We have already seen in Ibn Sina that the metals consist of the combination of sulphur and mercury. Even earlier the interaction of two parts were figuratively called impregnation. Both fuse into one symbol, and indeed so much the more readily, as it probably arose as the result of analogous thoughts, determined by a sexual complex. Also there occurs the idea that we must derive a male activity from the gold, a female from the silver, in order to get from their union that which perfects the mercury of the metals. That may be the reason that, for the above mentioned pair that

is to be united, the denotation gold and silver
( ⊙ and ☽ ) prevailed.   Red and white = man and
woman (male and female activity), we found in the
parable also when studied psychoanalytically.

In the " Turba philosophorum " " the woman is
called Magnesia, the white, the man is called red,
sulphur."

Morienus says.   " Our stone is like the creation
of man.   For first we have the union, 2, the corruption [i.e., the putrefaction of the seed], 3, the gestation, 4, the birth of the child, 5, the nutrition
follows."

Both constituents come from one root.   Therefore the authors inform us that the stone is an only
one.   If we call the matter " mercury," we therefore
generally speak of a doubled mercury that yet is
only one.

Arnold (Ros., II, 17) :   " So it clearly appears
that the philosophers spoke the truth about it, although it seems impossible to simpletons and fools,
that there was indeed only one stone, one medicine,
one regulation, one work, one vessel, both identical
with the white and red sulphur, and to be made at
the same time."

Id. (Ros., I, 6) :   " For there is only one stone,
one medicine, to which nothing foreign is added and
nothing taken away except that one separates the
superfluities from it."

Herein lies the idea of purification or washing; it
occurs again.   Arnold (Ros., II, 8) :   " Now when
you have separated the elements, then wash them."

The idea of washing is connected with that of mechanical purification, trituration, dismemberment in the parable, grinding (mill), and with the bath and solution (dissolution of the bridal pair). " Bath " is, on the other hand, the surrounding vessel, water bath. Arnold (Ros., I, 9) : " The true beginning, therefore, is the dissolution and solution of the stone." Fire can also cause a dissolution, either by fusion or by a trituration that is similar to calcination. They are all processes that put the substances in question into its purest or chemically most accessible form.

Arnold (Ros., I, 9) : " The philosophical work is to dissolve and melt the stone into its mercury, so that it is reduced and brought back to its prima materia, i.e., original condition, purest form."

Through the opening of the single substance the two things or seeds, red and white, are obtained.

But what is the " subject " that is put through these operations, the matter that must be so worked out? That is exactly what the alchemists most conceal. They give the prima materia (raw material) a hundred names, every one of which is a riddle. They give intimations of interpretations but are not willing to be definite. Only the worthy will find the keys to the whole work. The rest of the procedure can be understood only by one that knows the prima materia. Much is written on it and its puzzling names. They are, partly as raw material, partly as original material, partly as prime condition, called among other names Lapis philosophicus (philosopher's stone), aqua vitæ (water of life), venenum (poison), spiritus (spirit), medicina (medicine), cœlum (sky), nubes (clouds), ros (dew), umbra (shadow), stella signata (marked star), and Lucifer, Luna (moon), aqua ardens (fiery water), sponsa (betrothed), coniux (wife), mater, mother (Eve),— from her princes are born to the king,— virgo (virgin), lac virginis (virgin's milk), menstruum, materia hermaphrodita catholica Solis et Lunae (Catholic hermaphrodite matter of sun and moon), sputum Lunae (moon spittle), urina puerorum (children's urine), fæces dissolutæ (loose stool), fimus (muck), materia omnium formarum (material of all forms), Venus.

It will be evident to the psychoanalyst that the original material is occasionally identified with secretions and excretions, spittle, milk, dung, menstruum, urine. These correspond exactly to the infantile theories of procreation, as does the fact that these theories come to view where the phantasy forms symbols in its primitive activity. It is also to be noticed that countless alchemic scribblers who did not understand the works of the "masters" worked with substances like urine, semen, spittle, dung, blood, menstruum, etc., where the dim idea of a procreative essence in these things came into play. I will have something to say on this subject in connection with the Homunculus. I should meanwhile like to refer to the close relationship of excrement and gold in myth and folklore. [Cf. Note B at the end of this volume.] It is clear that for the art of gold production this mythological relationship is of importance.

To the action of analyzing substances before the reassembling or rebuilding, besides washing and trituration, belongs also putrefaction or rotting. Without this no fruitful work is possible. I have previously mentioned that it was thought that semen must rot in order to impregnate. The seed grain is subject to putrefaction in the earth. But we must remember also the impregnating activity of manure if we wish to understand correctly and genetically the association rot — procreate. Putrefaction is one of the forms of corruption (= breaking up) and corruptio unius est generatio alterius (the breaking up of one is the begetting of another).

Arnold (Ros., I, 9): "In so far as the substances here do not become incorporeal or volatile, so that there is no more substance [as such therefore destroyed] you will accomplish nothing in your work."

The red man and the white woman, called also red lions and white lilies, and many other names, are united and cooked together in a vessel, the philosophical Egg. The combined material becomes thereby gradually black (and is called raven or ravenhead), later white (swan); now a somewhat greater heat is applied and the substance is sublimated in the vessel (the swan flies up); on further heating a vivid play of colors appears (peacock tail or rainbow); finally the substance becomes red and that is the conclusion of the main work. The red substance is the philosopher's stone, called also our king, red lion, grand elixir, etc.  The after work is a subsequent elaboration by which the stone is given still more power, "multiplied" in its efficiency. Then in "projection" upon a baser metal it is able to tincture immense amounts of it to gold.  [In the stage of projection the red tincture is symbolized as a pelican.  The reason for this will be given later.] If the main work was interrupted at the white stage, instead of waiting for the red, then they got the white stone, the small elixir, with which the base metals can be turned into silver alone.

We have spoken just now of the main work and the after work.  I mention for completeness that the trituration and purification, etc., of the materials, which precedes the main work, is called the fore work.  The division is, however, given in other ways besides.

Armed with this explanation we can venture to look for the alchemic hieroglyphs in our parable.  I must beg the reader to recall the main episodes.

In the wanderer we have to conceive of a man who has started out to learn the secret of the great work. He finds in the forest contradictory opinions.  He has fallen deep into errors.  The study, although difficult, holds him fast.  He cannot turn back (Sec. 1).  So he pursues his aim still further (Sec. 2) and thinks he has now found the right authorities (Sec. 3) that can admit him to the college of wisdom. But the people are not at one with each other. They also employ figurative language that obscures the true doctrine, and which, contrasted with practice, is of no value. (I mention incidentally that the great masters of the hermetic art are accustomed to impress on the reader that he is not to cling to their words but measure things always according to nature and her possibilities.) The elders promise him indeed the revelation of important doctrines but are not willing to communicate the beginning of the work (Sec. 5, 6, preparation for the fight with the lion). That is a rather amusing trait of hermetic literature.

We have come to the fight with the lion, which takes place in a den. The wanderer kills the lion and takes out of him red blood and white bones, therefore red and white. Red and white enter later as roses, then as man and woman.

I cite now several passages from different alchemistic books.

Hohler (Herm. Phil., p. 91) says, apparently after Michael Meiers, " Septimana Philosophica ": " The green lion [a usual symbol for the material at the beginning] encloses the raw seeds, yellow hairs adorn his head [this detail is not lacking in the parable], i.e., when the projection on the metals takes place, they turn yellow, golden. [Green is the color of hope, of growth. Previously only the head of the lion is gold, his future. Later he becomes a red lion, the philosopher's stone, the king in robe of purple. At any rate he must first be killed.]

The lion that must die is the dragon, which the dragon fighter kills. Thus we have seen it in the mythological parallel. Psychoanalysis shows us further that lion = dragon = father (= parents, etc.). It is now very interesting that the alchemistic symbolism interchanges the same forms. We shall see that again.

Berthelot cites (Orig. de l'Alch., p. 60) from an old manuscript: "The dragon is the guardian of the temple. Sacrifice it, flay it, separate the flesh from the bones, and you will find what you seek."

The dragon is, as can be shown out of the old authors, also the snake that bites its own tail or which on the other hand can also be represented by two snakes.

Flamel writes on the hieroglyphic figure of two dragons (in the 3d chapter of his Auslegung d. hierogl. Fig.) the following: "Consider well these two dragons for they are the beginning of the philosophy [alchemy] which the sages have not dared to show their own children. . . . The first is called sulphur or the warm and dry. The other is called quicksilver or the cold and wet. These are the sun and the moon. These are snakes and dragons, which the ancient Egyptians painted in the form of a circle, each biting the other's tail, in order to teach that they spring of and from one thing [our lion!]. These are the dragons that the old poets represent as guarding sleeplessly the golden apples in the garden of the Hesperian maidens. These are the ones to which Jason, in his adventures of the golden fleece, gave the potion prepared for him by the beautiful Medea. [See my explanation of the

motive of dismemberment] of which discourses the
books of the philosophers are so full that there has
not been a single philosopher, from the true Hermes,
Trismegistus, Orpheus, Pythagoras, Artephias,
Morienus, and other followers up to my own time,
who has not written about these matters. These are
the two serpents sent by Juno (who is the metallic
nature) that were to be strangled by the strong Hercules (that is the sage in his cradle) [our wanderer],
that is to be conquered and killed in order to cause
them in the beginning of his work to rot, be destroyed and be born. These are the two serpents
that are fastened around the herald's staff and rod
of Mercury. . . . Therefore when these two (which
Avicenna calls the bitch of Carascene and the dog
of Armenia) are put together in the vessel of the
grave, they bite each other horribly. [See the battle of the sons of the dragon's teeth with Jason, the
elders in the parable, but also the embrace of the
bridal pair and the mythological parallels wrestling
= dragon fight = winning the king's daughter, . . .
= incest = love embrace or separation of the primal
parents, etc. . . . ". . . A corruption [destruc-
tion] and putrefaction must take place before the re-
newal in a better form. These are the two male and
female seed that are produced . . . in the kidneys
and intestines . . . of the four elements."

The dragon, who is killed at the beginning of the
work, is also called Osiris by the old alchemists.
We are now acquainted with his dismemberment,
also his relation to lead ore. Flamel calls the ves-

sel of the alchemistic operation a " grave." Olym-
piodorus speaks in an alchemistic work of the grave
of Osiris. Only the face of Osiris, apparently
wrapped up like a mummy, is visible. In the parable only the head of the lion is golden. The head
as the part preserved from the killing [dismember-
ment] stands probably for the organ of generation.
The phallus is indeed exactly what produces the procreating substance, semen. The phallus is the future. The phallus was consecrated by Isis as a
memorial.

Janus Lacinius gives in his Pretiosa Margarita
the following allegory. In the palace sits the king
decorated with the diadem and in his hand the
scepter of the whole world. Before him appears
his son with five servants and falling at his feet im-
plores him to give the kingdom to him and the servants. [The author takes the thing wrong end to.
The gold, king, is assailed by the other six metals,
because they themselves wish to be gold. The king
is killed. Essentially the same thing happens as
above.] Then the son in anger, and at the instigation of his companions, kills his father on the throne.
He collects the father's blood in his garment. A
grave [the lion's den, the grave] is dug, into which
the son intends to throw the father, but they both
fall in. [Cf. the dangerous walk of the wanderer
on the wall, Section 8, where the people fall off.]
The son makes every effort to get out again, but some
one comes who does not permit it. [Symbolism of
obstruction, the locked door, etc., in the parable.

The grave changes imperceptibly into the vessel where the bridal pair — with Lacinius they are father and son instead of mother and son — are united and securely locked in.] When the whole body is dissolved the bones are thrown out of the grave. They are divided into nine [dismemberment], the dissolved substance is cooked nine days over a gentle fire till the black appears. Again it is cooked nine days until the water is bright and clear. The black, with its water of life [in the parable the mill water is black] is cooked nine days till the white earth of the philosophers appears. An angel throws the bones on the purified and whitened earth, which is now mixed with its seeds. They are separated from water in a strong fire. Finally the earth of the bones becomes red like blood or ruby. Then the king rises from his grave full of the grace of God, quite celestial, with grand mien, to make all his servants kings. He places golden crowns on the heads of his son and the servants.

As bearers of both seeds, male and female, the lion is androgynous. Actually the subject (i.e., the first material) is conceived as twofold, bisexual. It is called by names that mean the two sexes, it is also called " hermaphrodite." It is represented as rebis (res bina = double thing), as a human with a male and a female head standing on a dragon. From the conquered dragon (lion) comes forth the Double. The substance is also called Mercurius; his staff bears the two antagonistic serpents mentioned by Flamel. In the parable also appears an hermaphrodite, the being (Sec. 8) which the wanderer cannot distinguish, whether it be a man or a woman. It is the original substance, Mercury, " our hermaphrodite."

In Section 9 of the parable, and also later, red and white appear in roses. The white and the red tincture are often in alchemy compared to white and red roses.

In Section 9 the wanderer comes to those houses where people work alone or by twos. They work in a slovenly fashion. The alchemistic quacks are generally called " bunglers " and " messy cooks " by the masters of the art. These are the ones who do not work according to the " possibilities of nature," which is, nevertheless, the touchstone of all right production.

The garden (Sec. 10, 11) is one of the " rose gardens " of which, e.g., the alchemist, Michael Meier, likes to speak.

There are difficulties in uniting the red youths with the white maidens. A wall separates them. The wanderer removes the obstruction in unlocking the door. That may indicate a chemical unlocking, by which the bodies are chemically brought nearer together.

The wanderer comes to a mill (Sec. 11). The mill naturally indicates the already mentioned trituration of the substance. It has, however, also reference to fermentation and in particular to that by means of meal.

Rulandus (Lex., pp. 211 ff., s. v. Fermentum) :

" Ferment is elixir, leaven, or yeast as it is called;
it makes porous the body that swells up and the
spirit finds a place in it so that it becomes fit to bake.
As now the meal is not yeast, but meal and water
[mill water] and the whole dough is thoroughly
leavened and real yeast, so also the lapis [stone] is
itself the ferment, yet gold and mercury are also
called ferment."

Now begins the main work — marriage, prison,
embrace, conception, birth, transfiguration — to
which the rest of the parable is devoted.

The prison is the philosophic egg. It is also
called " Athanor, a sieve, dunghill, bain-marie
(double cooker), a kiln, round ball, green lion,
prison, grave, brothel, vial, cucurbit." It is just like
the belly and the womb, containing in itself the true,
natural warmth (to give life to our young king).
The warmth that is used must first be gentle, "like
that after the winter "; it must be stronger like the
sun in spring, in summer [cf. the seasons in our para-
ble]. (Flamel, pp. 50 ff.)

Daustenius (Ros., VII) : ". . . And this thing
can be a symbol of a woman's belly, which, when
she has conceived, will immediately close the
womb."

Id. (Ros., VII) : " Therefore, when you have
put them (the white woman and the red man) in
their vessel, then close it as fast as possible. . . ."
[Seal of Hermes.]

Id. (Ros., VIII) : " Therefore that you arrange
the substances right and fine, and regulate your

work well, and marry consanguineous matter with masses acting consanguineously. . . ."  [Incest.]

Id. (Ros., VII) :  " So now this is our solution, that you marry the Gabricum with the Beja, which when he lies with the Beja, dies immediately and is changed into her nature.  Although the Beja is a woman, still she improves the Gabricum because he is come out of her."  [Death of the bridegroom son. It should be remembered in this connection that all metals or all substances generally — consequently also the ⊙ — come forth from the " mother," the primal substance ☿ .]

In a " Vision " of Daustenius, the king is to return into his mother's womb in order to be procreated afresh.  The king " goes into his bedroom and unexpectedly is fired with a great desire for coition, and goes to sleep at once, and has lain with a surpassingly beautiful maiden, who was a daughter of his mother" [weakened form of mother incest]. Later the vision says, " The woman, however, incloses her man, as a mother, quite carefully in the innermost part of her body."

The bodies inclosed in the vessel fall to pieces and are partly volatile.  The vapors [soul] return, however, into the bodies.  There conception takes place.

Daustenius [Ros. IX.] : ". . . From that are airy spirits come, that with each other rise into the air, and there have conceived life, that is blown into them by their dampness, as the human being has life from air, by which it increases . . . For life of all natural things depends upon the blowing in of air."

The bestowing of life by a blowing in of air plays a great part in myths. Also there occurs quite frequently special impregnations by air and wind. It is a primitive impregnation theory, that is found also in the ideas of children.

Children think of the blowing in of air into the anus as a natural sexual theory. I know several cases where this practice is carried out with emphasis on the erotic under the pretense of " playing doctor." A child once told what papa and mamma do when they are alone; they put their naked backsides together and blow air into each other.

Another infantile theory explains impregnation by the swallowing of an object. In myths and fairy lore this motive occurs with extraordinary frequency. To the swallowing as conception, corresponds defecation as parturition. Incidentally we should note that the bodies in the philosophic egg turn actually into a rolling, stinking, black mass, which is expressly called dung by many authors. The water is also called urine. The prima materia is also called urine. In the philosophical egg the white woman swallows the red man, man-eating motive. (Stucken.)

Liber Apocal. Hermetis (Cited by Hohler, p. 105 f.) : " . . . Therefore the philosophers have married this tender young maiden to Gabricus, to have them procreate fruit, and when Gabricus sleeps he dies. The Beja [i.e., the white maiden] has swallowed him and consumed him because of her great love."

Now as to the intra-uterine nourishment of the fetus by means of the water of life:

Daustenius [Ros. VI.]: " . . . The fruit in the womb is nourished only by the mother's blood." Id. (Ros. x) :

" Without seeds no fruit can grow up for thee:
First the seed dies; then wilt thou see fruit.
In the stomach the food is cooked tender
From which the limbs draw the best to themselves.
When too the seed is poured into the womb
Then the womb stays right tenderly closed.
The menstruum does not fail the fruit for nourishment
Till it at the proper time comes to the light of day."

Later he says (Id., XI) : " Lay the son by her that she suckle him." [The water of life is therefore also the milk.]

The new king is born, and now he and his consort appear in priceless garments (cf. Section 18 of the parable). The color change of the substance is expressed by means of the change of garments, like peacock's tail, rainbow. The process goes from black through gray to white, yellow, red, purple.

The end is reached with purple. The wanderer at the end describes the virtues of the philosopher's stone. We have already compared the great elixir with soma. In the old alchemistic book, which bears the name of the Persian magician, Osthanes (Berthelot, Orig., p. 52), the divine water heals all maladies. Water of life,— elixir of life.

Many readers will shake their heads over the psychoanalytic exposition of the parable. The gross development of sexuality and the Œdipus complex may seem improbable to him. The alchemistic hieroglyphic has now in unexpected manner shown after all, that these surprising things were not read into the parable by psychoanalysis, but rightly found in it, even though psychoanalysis has not by any means exhausted the contents of the parable. What might at first have appeared to be bold conjecture, as for example, killing of the father, incest with the mother, the conception of the red blood and white bones as man and woman, the excrementitious substance as procreative, the prison as the uterus, has all been shown to be in use as favorite figurative expression among the alchemistic authors.

The alchemists like to dwell on the process of procreation, and on infantile sexual theories. The deep interest that they show in these matters, and without which they would not have used them so much in their hieroglyphics, the meaning that these things must have, in order to be regarded as worthy to illustrate the processes of the great work, and finally, the meaning that in some form or other they actually have in the emotional life of every man, all of this makes it evident that the line of imaginative speculations with which we have become acquainted, deserves independent treatment. In practice there was a fission, and procreation becomes an independent problem for alchemists. Yet the followers of the art did learn from nature, in order that their art might follow the works of nature even to improve on her; what wonder then if many of them set themselves to the artificial creation — generation — of man? Yet the belief in generatio equivoca has not long been dead. Must it not have seemed somehow possible, in view of the supposed fact that they saw insects develop out of earth, worms out of dung, etc., that they should by special artificial interposition, be able to make higher forms of life come out of lifeless matter? And of all the substances not one was indeed completely lifeless for the " animated " metals even, grew and increased. In short, if we regard the matter somewhat more closely, it is after all not so extraordinary that they made serious attempts to create the homunculus.

Generally Paracelsus is regarded as the author of the idea, which to the somewhat uncritical, could not, in my opinion, help being in the air. There are different views regarding the part played by Paracelsus. The instructions that he gives for the production of the homunculus are found in a work (De natura rerum) whose authorship is not settled. And supposing that Paracelsus was the writer, it must be considered whether he does not lay before the inquisitive friend to whom the work is dedicated merely a medley of oddities from the variegated store that he had collected from all sources on his travels among vagrant folk. We must accept the facts as we find them; the question as to whether it was Paracelsus or not would be idle. Enough that there is a book by some writer who describes the work and describes it in such a way that naïve scholarship could have thought it quite consistent. The idea as such has appeared conceivable to us. Its form in the book mentioned appears clearly determined by alchemistic ideas. The reader will immediately perceive it himself as I give here some passages from the book. (Cf. the Strassb. Folio Ausg. des Paracelsus, Vol. I, pp. 881–884.) A consideration of the production of the homunculus appears important to me because it shows the main content of alchemistic ideas in enlarged form and complete development, a content that gives, moreover, the very thing that psychoanalysis would here look for.

Paracelsus begins with the fact that putrefaction transforms all things into their first shape and is the beginning of generation and multiplication. The spagiric [One of the names for alchemy. From σπᾶν (separate), and ἀγείρειν (unite).] art is able to create men and monsters. Such a monster is the Basilisk. " The Basilisk " grows and is born out of and from the greatest impurity of women, namely from the menstrua and from the blood of sperm that is put into a glass and cucurbit, and putrefied in a horse's belly. In such putrefaction is the Basilisk born. Whoever is so daring and so fortunate as to make it or to take it out or again to kill it, who does not clothe and protect himself before with mirrors? I advise no one but I wish to give sufficient warning. [Many fables about the Basilisk were then current. The belief, too, was general that this terrible animal was produced from a hen's egg.

Herein lies, again, the idea of unnatural procreation.] . . . Now the generation of the homunculus is not to be forgotten.  For there is something in it, notwithstanding that it has till now been kept in mystery and concealed, and that not a little doubt and question there was among some of the old philosophers, whether it was possible for art and nature that a man should be born outside a woman's body and a natural mother.  To which I give the answer that it is in no way contrary to the spagiric art and nature, but is quite possible; but how such accomplishment and occurrence may be, is by the following procedure:  Namely that the semen of a man is putrefied in a closed cucurbit per se, with the greatest putrefaction in a horse's belly for 40 days or until it comes to life and moves and stirs, which is easily to be seen.  [Horse's belly by metonomy for horse's dung.  Horse manure or dung was an easily procured material that served the purpose of keeping warm at an even mild and moist heat a vessel that was put into it.  Horse manure is then finally the gentle " moist heat " in general engendered by any means.  In the preceding case surely the narrower meaning of animal belly or dung should not be overlooked.  Here indeed this belly with its moist warmth has to act as an equivalent for a uterus.]  After such a time it will look something like a man but transparent without a body.  So after this it is daily fed whitish (weisslich) with the Arcano sanguinis humani [the water of life that nourishes the fœtus] and nourished about 40 weeks and kept in the even warmth of a horse's belly. A real live human child will come forth with all members like another child that is born of a woman but much smaller. We call it homunculus and it should then be brought up just like another child with great diligence and care till it comes to its days of understanding. That is now the highest and greatest mystery that God has let mortal and sinful man know. For it is a miracle and magnale Dei, and a mystery above all mystery and should also be kept a mystery fairly till the judgment day, as then nothing will stay hidden, but all will be revealed.

" And although such a thing has hitherto been hidden from natural man, it has not been hidden from the fauns and the nymphs and giants, but has been revealed for a long time; whence they too, come. For from such homunculi, when they come to the age of manhood come giants, dwarfs and other similar great wonder people, [Just like Genesis VI, 4] that were used for a great tool and instrument, who had a great mighty victory over their enemies and knew all secret and hidden things that are for all men impossible to know. For by art they received their life, through art they received body, flesh, bone and blood, through art were they born. Therefore the art was embodied and born in them and they had to learn it from no one, but one must learn from them. For because of art are they there and grown up like a rose or flower in the garden and are called the children of fauns and nymphs because that they with their powers and deeds, not to men but to spirits are compared." [It is characteristic that Paracelsus passes immediately to the production of metals.]

In the description of the generation of the homunculus the power of rotting material has been pointed out. There is clearly evident a feeding with a magisterium from blood (water of life) corresponding to the intrauterine alimentation. We note that from the homunculi come giants and dwarfs and wonderful beings.

The idea of palingenesis appears to have no little significance for the existence of the homunculus production. They imagine that a dead living being could be restored, at least in a smokelike image, if they carefully collected all its parts, triturated them and treated the composition in a vessel with the proper fire. Then there would appear after a time, like a cloud of smoke, the faint image of the former being, plant, bird, man. The clouds vanish if the heating is interrupted. Further it would be possible, even if more difficult, to pass beyond this mere adumbration, and cause the former being to arise again from the ashes, fully alive. In the recipes for this an important rôle is regularly played by horse manure or some other rotting substance. Many authors tell fables of all sorts of wonderful experiments that they have made. One tells that he has reduced a bird to ashes and made it live again, another will have seen in his retort and coming from the moldering corpse of a child its shadow image, etc. We see here in actuality the mythical motive of dismemberment and revivification expressed in a naïve practice. It is quite noticeable that this practice follows the same lines as the mythical representation. *All* the constituent parts of the body that is *cut into little pieces* must be carefully collected and put in a *vessel* and (generally) *cooked.*

The human child as result of cooking or else of a similar process in a vessel, is not infrequent in primitive myths. I could mention a Zulu myth (Frobenius, Zeitalt. d. Sonneng., I, p. 237) of a formerly barren woman. It was said that she should catch a drop of blood in a pot, cover it up and set it by for eight months, and should open it in the ninth month. The woman did as she was advised and found a child in the pot. The drop of blood, be it noted, came from herself. The numerous whale dragon myths (Frobenius) where it is very hot inside of the whale, belong here in motive. From the whale's belly comes indeed the baked young (sun) hero. [Who moreover generally gets nourishment in the whale-dragon's belly. Nutritio. Heart motive according to Frobenius.] It is interesting that the idea of cooking human beings occurs very clearly in a well analyzed case of dementia precox. (Spielrein in Jb. ps. F., III, pp. 358 ff.) In the strongly regressive phantasies of the invalid, fragments of all sorts of things are cooked or roasted and the ashes can become men."

A very interesting variant of the infantile theories of procreation of the living in dung is found in the book, " De Homunculis et Monstris " (Vol. II, pp. 278 ff. of the Strassburg edition of the works of

Paracelsus). It is there maintained that by sodomy as well as by pederasty (specifically coitus in anum and also in os is meant) the generation of a monster is possible.

As they did with alchemy in general, so charlatans also made use of the production of the homunculus. Their business was based on the great profits that were offered by the possession of a homunculus and that are equivalent to those of mandrake alum. Mandrake alum gave a certain impetus to the development of the homunculus idea and practice. It can be shown that secrets of procreation seem partly to underlie this also.

It is easy to show the possibility that many a duffer was led toward the production of the homunculus by erroneous interpretation of the procreation symbolism occurring in the alchemistic writings. It was merely necessary, in their limitations, to take literally one or another of the methods. In this way there actually occurred the most ludicrous blunders. Because the philosopher's egg was mentioned, they took eggs as the actual subject. Because the spermatic substance and seeds were mentioned they thought that the prima materia was human semen, and so arose the school of seminalists. And because it was written of the subject that it was to be found wherever men dwell, and that it was a little despised thing which men threw away not realizing its worth, and because they thought of putrefaction as such, they thought to find the real substance in human excrement, and so the school of stercoralists was

founded. From the belief in the healing and won-
derworking power of excrement sprang moreover
the famous filth pharmacy, that was held in no little
esteem.

The homunculus topic is exceedingly interesting.
Unfortunately I cannot in the space of this book go
into it thoroughly. I shall do so in another place.

## SECTION III

## THE HERMETIC ART

ANY one that makes a thorough study of the alchemistic literature must be struck with the religious seriousness that prevails in the writings of the more important authors. Every "master" who enjoyed the highest honor among his fellows in the hermetic art has a certain lofty manner that keeps aloof from the detailed description of chemical laboratory work, although they do not depart from the alchemistic technical language. They obviously have a leaning toward some themes that are far more important than the production of a chemical preparation can be, even if this is a tincture with which they can tinge lead into gold. Looking forth to higher nobler things, these authors, whose homely language frequently touches our feelings deeply, make the reader notice that they have nothing in common with the sloppy cooks who boil their pots in chemical kitchens, and that the gold they write about is not the gold of the multitude; not the venal gold that they can exchange for money. Their language seems to sound as if they said, "Our gold is not of this world." Indeed they use expressions that can with absolute clearness be shown to have this sense. Authors of this type did not weary of enjoining on the

novices of the art, that belief, scripture and righteousness were the most important requisites for the
alchemistic process. [With the sloppers it was indeed a prime question, how many and what kinds of
stoves, retorts, kettles, crucibles, ores, fires, etc., in
short, what necessary implements they needed, for
the great work.]

He whose eyes are open needs no special hints to
see, in reading, that the so-called alchemistic pre-
scriptions did not center upon a chemical process. A
faint notion of the circumstance that even in their
beginnings, alchemistic theories were blended with
cosmogonic and religious ideas, must make it quite
evident that, for example, in the famous Smaragdine
Tablet of Hermes [Its real author is unknown.] a
noble pillar of alchemy, something more must be
contained than a mere chemical recipe. The language of the Smaragdine tablet is notoriously the
most obscure that the hermetic literature has produced; in it there are no clear recommendations to
belief or righteousness; and yet I think that an unprejudiced reader, who was not looking specially
for a chemical prescription, would perceive at least
a feeling for something of philosophy or theology.

[1.] Verum, sine mendacio, certum et veris-
simum: [2.] Quod est inferius est sicut quod est.
superius, et quod est superius est sicut quod est inferius, ad perpetranda [also: penetranda, praepa-
randa] miracula rei unius. [3.] Et sicut res omnes
fuerunt ab uno, meditatione unius: sic omnes res
natae fuerunt ab hac una re, adaptatione [adop-

tione]. [4.] Pater ejus est Sol, mater ejus est
Luna. [5.] Portavit illud ventus in ventre suo.
[6.] Nutrix ejus terra est. [7.] Pater omnis Te-
lesmi totius mundi est hic. [8.] Virtus ejus integra
est, si versa fuerit in terram. [9.] Separabis terram ab igne, subtile ab spisso, suaviter, magno cum
ingenio. [10.] Ascendit a terra in coelum, ite-
rumque descendit in terram, et recipit vim supe-
riorum et inferiorum. [11.] Sic habebis Gloriam
totius mundi. Ideo fugiet a te omnis obscuritas.
Haec est totius fortitudinis fortitudo fortis, quia
vincet omnem rem subtilem, omnemque solidam
[solidum] penetrabit. [12.] Sic mundus creatus
est. [13.] Hinc erunt adaptationes mirabiles,
quarum modus est hic. [14.] Itaque vocatus sum
Hermes Trismegistus, habens tres partes philosophiae totius mundi. [15.] Completum est quod
dixi de operatione Solis.

Translation: [1.] It is true, without lies and
quite certain. [2.] What is lower is just like what
is higher, and what is higher is just like what is
lower, for the accomplishment of the miracle of a
thing. [3.] And just as all things come from one
and by mediation of one, thus all things have been
derived from this one thing by adoption. [4.] The
father of it is the sun, the mother is the moon. [5.]
The wind has carried it in his belly. [6.] The earth
has nourished it. [7.] It is the father [cause] of
all completion of the whole world. [8.] His
power is undiminished, if it has been turned towards
the earth. [9.] You will separate the earth from

fire, the fine from the coarse, gently and with great skill. [10.] It ascends from the earth to the sky, again descends to the earth, and receives the powers of what is higher and what is lower. [11.] Thus you will have the glory of the whole world, and all darkness will depart from you. It is the strength of all strength, because it will conquer all the fine and penetrate all the solid. [12.] Thus the world was created. [13.] From this will be wonderful applications of which it is the pattern. [14.] And so I have been called Hermes, thrice greatest, possessing three parts of the knowledge of the whole world. [15.] Finished is what I have said about the work of the sun.

Sun and gold are identical in the hieroglyphic mode of expression. Whoever seeks only the chemical must therefore read: The work of gold, the production of gold; and that is what thousands and millions have read. The mere word gold was enough to make countless souls blind to everything besides the gold recipe that might be found in the Smaragdine tablet. But surely there were alchemistic masters who did not let themselves be blinded by the word gold and sympathetically carried out still further the language of the Smaragdine tablet. They were the previously mentioned lofty-minded men. The covetous crowd of sloppers, however, adhered to the gold of the Smaragdine tablet and other writings and had no appreciation of anything else. For a long time alchemy meant no more for modern historians.

The fact that modern chemical science is sprung from the hermetic works,— as the only branch at present clearly visible and comprehensible of this misty tree of knowledge,— has had for result that in looking back we have received a false impression. Chemical specialists have made researches in the hermetic art and have been caught just as completely in the tangle of its hieroglyphics as were the blind seekers of gold before them.   The hermetic art, or alchemy in the wider sense, is not exclusively limited to gold making or even to primitive chemistry.   It should, however, not be surprising to us who are acquainted with the philosophical presuppositions of alchemy, that in addition to the chemical and me- chanical side of alchemy a philosophical and religious side also received consideration and care.   I think, however, that such historical knowledge was not at all necessary to enable us to gather their pious views from the religious language of many masters of the hermetic art.   However, this naïve childish logic was a closed book to the chemists who made histori- cal researches.   They were hindered by their special knowledge.   It is far from my purpose to desire in the least to minimize the services that a Chevreul or a Kopp has performed for the history of chemis- try; what I should like to draw attention to is merely that the honored fathers of the history of chemistry saw only the lower —" inferius "— and not the higher —" superius "— phase of alchemy, for exam- ple, in the Smaragdine tablet; and that they used it as the type of universal judgment in such a way that it needed a special faculty for discovery to reopen a fountain that had been choked up.

I now realize that the poets have been more fortunate than the scientists.   Thus Wieland, who, for example, makes Theophron say in the Musarion (Book II) :

> The beautiful alone
> Can be the object of our love.
> The greatest art is only to separate it from its tissue . . .
> For it [the soul] nothing mortal suffices,
> Yea, the pleasure of the gods cannot diminish a thirst
> That only the fountain quenches.  So my friends
> That which other mortals lures like a fly on the hook
> To sweet destruction
> Because of a lack of higher discriminative art
> Becomes for the truly wise
> A Pegasus to supramundane travel.

But the poets usually speak only in figures.  I will therefore rest satisfied with this one example.

The service of having rediscovered the intrinsic value of alchemy over and above its chemical and physical phase, is to be ascribed probably to the American, Ethan Allen Hitchcock, who published his views on the alchemists in the book, " Remarks upon Alchemy and the Alchemists," that appeared in Boston in 1857, and to the Frenchman, N. Landur, a writer on the scientific periodical " L'Institut," who wrote in 1868 in similar vein [in the organ " L'Institut," 1st Section, Vol. XXXVI, pp. 273 ff.], though I do not know whether he wrote with knowledge of the American work.   Landur's observations are reported by Kopp (Alch., II, p. 192), but he does not rightly value their worth. It need not be a reproach to him. He undertook as a chemical specialist a work that would have required quite as much a psychologist, a philosopher or a theologian.

The discoveries made by the acute Hitchcock are so important for our analysis, that a complete exposition of them cannot be dispensed with. I should like better to refer to Hitchcock's book if it were not practically inaccessible.

We have heard that the greatest stumbling block for the uninitiated into the hermetic art lay in the determination of the true subject, the prima materia. The authors mentioned it by a hundred names; and the gold seeking toilers were therefore misled in a hundred ways. Hitchcock with a single word furnishes us the key to the understanding of the hermetic masters, when he says: The subject is man. We can also avail ourselves of a play on words and say the subject or substance is the subject.

The uninitiated read with amazement in many alchemists that "our subjectum," that is, the material to be worked upon, is also identical with the vessel, the still, the philosopher's egg, etc. That becomes intelligible now. Hitchcock writes (H. A., p. 117) very pertinently: "The work of the alchemists was one of contemplation and not a work of the hands. Their alembic, furnace, cucurbit, retort, philosophical egg, etc., etc., in which the work of fermentation, distillation, extraction of essences and spirits and the preparation of salts is said to have taken place was Man,— yourself, friendly reader,— and if you will take yourself into your own study and be candid and honest, acknowledging no other guide or authority but Truth, you may easily discover something of hermetic philosophy; and if at the beginning there should be ' fear and trembling ' the end may be a more than compensating peace."

The alchemist Alipili (H. A., p. 34) writes: " The highest wisdom consists in this, for man to know himself, because in him God has placed his eternal Word. . . . Therefore let the high inquirers and searchers into the deep mysteries of nature learn first to know what they have in themselves, and by the divine power within them let them first heal themselves and transmute their own souls, . . . if that which thou seekest thou findest not within thee, thou wilt never find it without thee.   If thou knowest not the excellency of thine house, why dost thou seek and search after the excellency of other things?   The universal Orb of the world contains not so great mysteries and excellences as does a little man formed by God in his own image.   And he who desires the primacy amongst the students of nature, will nowhere find a greater or better field of study than himself.   Therefore will I here follow the example of the Egyptians and . . . from certain true experience proclaim, O Man, know thyself; in thee is hid the treasure of treasures."

A seminalist has concluded from this that the

prima materia is semen, a stercoralist, that it is dung.

George Ripley describes the subject of the philosopher's stone as follows:

" For as of one mass was made the thing,
Right must it so in our praxis be,
All our secrets of one image must spring;
In philosophers' books therefore who wishes may see,
Our stone is called the less-world, one and three."

The stone is therefore the world in little, the microcosm, man; one, a unity, three, ☿ mercury, ♁ sulphur, ⊖ salt, or spirit, soul, body. Dichotomy also appears, mercury and sulphur, which can then generally be rendered soul and body. One author says, " We must choose such minerals as consist of a living mercury and a living sulphur; work it gently, not with haste and hurry." [Cf. Tabula Smaragdina 9, " suaviter " . . .]

Hitchcock (H. A., p. 42): " The ' one ' thing of the alchemists is above all man, according to his nature [as a nature] essentially and substantially one. But if the authors refer to man phenomenally they speak of him under different names, indicating different states as he is before or after ' purification ' or they refer to his body, his soul or his spirit under different names. Sometimes they speak of the whole man as mercury, . . . and then by the same word perhaps they speak of something special, as our mercury which has besides, a multitude of other names . . . although men are of diverse dispositions and temperaments, some being angelic and others satanic, yet the alchemists maintain with St. Paul that ' all the nations of men are of one blood,' that is, of one nature. And it is that in man by which he is of one nature which it is the special object of alchemy to bring into life and activity; that by whose means, if it could universally prevail, mankind would be constituted into a brotherhood."

The alchemist says that a great difficulty at the outset of the work is the finding or making of their necessarily indispensable mercury, which they also call green lion, mercurius animatus, the serpent, the dragon, acid water, vinegar, etc.

What is this mysterious mercury, susceptible to evolution, lying in mankind, common to all, but differently worked out? Hitchcock answers, conscience. Conscience is not equally " pure " with all men, and not equally developed; the difficulty of discovering it, of which the alchemists tell, is the difficulty of arousing it in the heart of man for the heart's improvement and elevation. The starting point in the education of man is indeed to awaken in his heart an enduring, permanent sense of the absolutely right, and the consistent purpose of adhering to this sense. It is above all one of the hardest things in the world " to take a man in what is called his natural state, St. Paul's natural man, after he has been for years in the indulgence of all his passions, having a view to the world, to honors, pleasures, wealth, and make him sensible of the mere abstract claims of right, and willing to relinquish one single passion in deference to it." Surely that is the one great task of the educator; if it be accomplished, the work of improvement is easy and can properly be called mere child's play, as the hermetics like to call the later phases of their work.    (H. A., pp. 45 ff.)

No one is so suspicious and so sensitive as those whose conscience is not sensitive enough.    Such people who wander in error themselves, are like porcupines: it is very difficult to approach them.    The alchemists have suitable names for them as arsenic, vipers, etc., and yet they seek in all these substances, and in antimony, lead, and many other materials, for a true mercury that has just as many names as there are substances in which it is found; oil, vinegar, honey, wormwood, etc.    Under all its names mercury is still, however, a single immutable thing. It was also called an incombustible sulphur for whoever has his conscience once rightly awakened, has in his heart an endlessly burning flame that eats up everything that is contrary to his nature.    This fire that can burn like " poison " is a powerful medicine, the only right one for a (morally) sick soul.

Conscience in the crude state is generally called by the alchemists " common quicksilver " in contrast to " our quicksilver."    To replace the first by the second and, according to the demands of nature, not forcibly, is the one great aim that the hermetics follow.    This first goal is a preparation for a further work.    Whither this leads we can represent in one word —" God "— and even here we may be struck with the " circular " character of the whole hermetic work, since the heavenly mercury that is necessary

to the preliminary work, to the purification, is yet
itself a gift of God; the beginning depends on the
end and presupposes it.   The symbol of the prima
materia is not without purpose a snake that has its
tail in its mouth.   I cannot, in anticipation, enter
into the problem that arises in this connection; only
let it be understood in a word that the end can soar
beyond the beginning as an ideal.

What is to be done with the messenger of heaven,
mercury, or conscience, when it has been discovered?
Several alchemists give the instruction to sow the
gold in mercury as in the earth, " philosophic gold "
that is also called Venus-love.   Often the New Testament proves the best commentary on the hermetic
writings.   In Corinthians III, 9, ff., we read:
" Ye are God's husbandry, ye are God's building.
According to the grace of God which is given unto
me as a wise master builder, I have laid the foundation and another buildeth thereupon.   For other
foundation can no man lay than that is laid, which
is Jesus Christ.   Now if any man build upon this
foundation gold, silver, precious stones, hay, stub-
ble. . . . Every man's work shall be made manifest
. . . because it shall be revealed by fire; and the fire
shall try every man's work of what sort it is."   And
Galatians VI, 7 ff.:   " For whatsoever a man soweth,
that shall he also reap.   For he that soweth to his
flesh shall of the flesh reap corruption; but he that
soweth to the Spirit shall of the Spirit reap life everlasting.   And let him not be weary in well doing; for
in due season we shall reap if we faint not."   The

spirit to which it is sowed there is ☿, mercury, and
the gold that will come out is to be proved in the
fire.

The alchemists speak of men very often as of
metals.   Before I cite from the work of Johann
Isaak Hollandus on lead, I call to mind that lead, ♄,
bears the name of Saturn.   The writing of Hollandus could quite as well be called a treatise on mankind as on lead.   To understand this better, be it
added that man in a state of humility or resignation
must specially be associated with lead, the soft, dark
metal.

The publisher of the English translation of J. I.
Hollandus, which is dated 1670, addresses the
reader as follows:   " Kind reader, the philosophers
have written much about their lead, which as Basilus
has taught, is prepared from antimony; and I am
under the impression that this saturnine work of the
present philosopher, Mr. Johann Isaak Hollandus,
is not to be understood of common lead . . . but of
the lead of the philosophers."

And in Hollandus himself we read:   " In the
name of God, Amen.—   My child, know that the
stone called the Philosopher's Stone comes from
Saturn.   And know my child as a truth that in the
whole vegetable work [vegetable on account of the
symbolism of the sowing and growing] there is no
higher or greater secret than in Saturn.   [Cf. the
previously cited passage from Alipili.]   For we find,
ourselves, in [common] gold not the perfection that
is to be found in Saturn, for inwardly he is good

gold. In this all philosophers agree; and it is necessary only that you reject everything that is superfluous, then that you turn the within outward, which is the red; then it will be good gold. [H. A., p. 74, notes that Hollandus himself means the same as Isaiah L, 16. "Wash you, make you clean; put away the evil of your doings from before mine eyes," etc.] Gold cannot be made so easily of anything as of Saturn, for Saturn is easily dissolved and congealed, and its mercury may be more easily extracted from it." [That means therefore that the conscience easily develops after the destruction of superfluities or obstacles in the plastic lead man.] "And this mercury extracted from Saturn is purified and sublimated, as mercury is usually sublimed. I tell thee, my child, that the same mercury is as good as the mercury extracted from gold in all operations." [Herein lies, according to H. A., an allusion to the fact that all men are essentially of one nature, inasmuch as the image of God dwells in them all.]

"All these strange parables, in which the philosophers have spoken of a stone, a moon, a stove, a vessel, all of that is Saturn [i.e., all of that is spoken of mankind] for you may add nothing foreign, outside of what springs from himself. There is none so poor in this world that he cannot operate and promote this work. For Luna may be easily made of Saturn in a short time [here Luna, silver, stands for the affections purified]; and in a little time longer Sol may be made from it. By Sol here I understand the intellect, which becomes clarified in propor-

tion as the affections become purified. . . . In
Saturn is a perfect mercury; in it are all the colors
of the world, [that is, the whole universe in some
sense lies in the nature of man, whence have proceeded all religions, all philosophies, all histories, all
fables, all poesy, all arts and sciences.]"    (P. 77.)

Artephius [Hapso]:  "Without the antimonial
vinegar [conscience] no metal [man] can be whitened [inwardly pure]. . . . This water is the only
apt and natural medium, clear as fine silver, by which
we ought to receive the tinctures of Sol and Luna
[briefly, if also inexactly, to be paraphrased by soul
and body], so that they may be congealed and
changed into a white and living earth."   This water
desires the complete bodies in order that after their
dissolution it may be congealed, fixed and coagulated
into a white earth.   [The first step is purification,
releasing, that is, otherwise also conceived as calcination, etc.; it takes place through conscience, under
whose influence the hard man is made tender and
brought to fluidity.]

"But their [sc. the alchemists] solution is also
their coagulation; both consist in one operation, for
the one is dissolved and the other congealed.   Nor
is there any other water which can dissolve the bodies
but that which abideth with them.   Gold and silver
[Sol and Luna as before] are to be exalted in our
water, . . . which water is called the middle of the
soul and without which nothing can be done in our
art.   It is a vegetable, mineral, and animal fire,

which conserves the fixed spirits of Sol and Luna,
but destroys and conquers their bodies; for it an-
nihilates, overturns and changes bodies and metallic
forms, making them to be no bodies, but a fixed
spirit."

" The argentum vivum [living silver] is . . . the
substance of Sol and Luna, or silver and gold,
changed from baseness to nobility.

" It is a living water that comes to moisten the
earth that it may spring forth and in due season
bring forth much fruit. . . . This aqua vitæ or
water of life, whitens the body and changes it into
a white color. . . .

" How precious and how great a thing is this
water.   For without it the work could never be
done or perfected; it is also called vas naturae, the
belly, the womb, receptacle of the tincture, the earth,
the nurse.   It is the royal fountain, in which the
king and queen [☉ and ☽] bathe themselves; and
the mother, which must be put into and sealed up
within the belly of her infant, and that is Sol himself,
who proceeded from her, and whom she brought
forth; and therefore they have loved one another as
mother and son, and are conjoined together because
they sprang from one root and are of the same substance and nature.   And because this water is the
water of the vegetable life, it causes the dead body to
vegetate, increase and spring forth, and to rise from
death to life, by being dissolved first and then sublimed.   And in doing this the body is converted into

a spirit and the spirit afterwards into a body. . . .

"Our stone consists of a body, a soul, and a spirit.

"It appears then that this composition is not a work of the hands but a change of natures, because nature dissolves and joins itself, sublimes and lifts itself up, and grows white being separated from the feces [these feces are naturally the same that Hollandus notes as the "superfluities"]. . . . Our brass or latten then is made to ascend by the degrees of fire, but of its own accord freely and without violence. But when it ascends on high it is born in the air or spirit and is changed into a spirit, and becomes a life with life. And by such an operation the body becomes of a subtile nature and the spirit is incorporated with the body, and made one with it, and by such a sublimation, conjunction and raising up, the whole, body and spirit, is made white." (H. A., p. 87.)

For elucidation some passages from the Bible may be useful. Colossians II, 11: "In whom also ye are circumcised with the circumcision made without hands, in putting off the body of the sins of the flesh by the circumcision of Christ." Psalm LI, 7: "Wash me and I shall be whiter than snow." I Corinthians VI, 11: "But ye are washed, but ye are sanctified, but ye are justified in the name of the Lord Jesus." Romans VIII, 13: "For if ye live after the flesh, ye shall die; but if ye through the Spirit do mortify the deeds of the body, ye shall live." John IV, 14: "But whosoever drinketh of the water that I shall give him, shall never thirst; but the water that I shall give him shall be in him a well of water springing up into everlasting life." [In IV, 10, living water is mentioned.]    John XII, 24 ff.: ". . . Except a corn of wheat fall into the ground and die [Putrefactio] it abideth alone; but if it die it bringeth forth much fruit.    He that loveth his life shall lose it and he that hateth his life in the world shall keep it unto life eternal."

Romans VI, 5 ff.: " For if we have been planted together in the likeness of his death, we shall be also in the likeness of his resurrection.    Knowing this, that our old man is crucified with him [I must mention here that the hieroglyph for vinegar is ✠] that the body of sin might be destroyed. . . ."

I Corinthians XV, 42 ff.: " It is sown in corruption, it is raised in incorruption.    It is sown in dishonor, it is raised in glory. . . . It is sown a natural body, it is raised a spiritual body. . . . The first Adam was made a living soul, the last Adam was made a quickening spirit. . . . We shall all be changed. . . . For this corruptible must put on incorruption, and this mortal must put on immortality."

I Corinthians XV, 40 ff.: " There are celestial bodies and bodies terrestrial. . . . There is one glory of the sun and another glory of the moon."

Ephesians II, 14 ff.: " For he is our peace, who hath made both one, and hath broken down the middle wall of partition between us, having abolished in his flesh the enmity, even the law of commandments contained in ordinances; for to make in himself of twain one new man, so making peace, and that he might reconcile both unto God in one body by the cross, having slain the enmity thereby."

If we note the two contraries that are to be united according to the procedure of the hermetic philosophers with ☉ and ☽ [sun and moon, gold and silver, etc.] and represent them united with the cross ✛ we get ☿̄ ; i.e., ☿ , the symbol of mercury. This ideogram conceals the concept, Easter. All these ideas, as we know, did not originate with Christianity.

II Corinthians v, 1 : " For we know that if our earthly house of this tabernacle were dissolved, we have a building of God, an house not made with hands, eternal in the heavens."

John VII, 38 : " He that believeth on me . . . out of his belly shall flow rivers of living water."

I mention right here that the hermetic philosophers do not pursue speculative theology, but that, as is clearly evident from their writings, they made the content of the religious doctrine a part of their life. That was their work, a work of mysticism. Everything that the reader is inclined to conceive in the passages above, as probably belonging merely to the other life, they as Mystics, sought to represent to themselves on earth, though without prejudice to the hope of a life beyond. I presume that they therefore speak of two stones, a celestial and a terrestrial. The celestial stone is the eternal blessedness and, as far as the Christian world of ideas is considered, is Christ, who has aided mankind to attain it.   The terrestrial stone is the mystical Christ whom each may cause to be crucified and resurrected in himself, whereby he attains a kingdom of heaven on earth with those peculiar qualities that have been allegorically attributed to the philosopher's stone. Therefore the terrestrial stone is called a reflection of the celestial and so it is said that from lead, etc., the stone may be easily produced and " in a short time," i.e., not only after death.

At any rate in primitive symbolism there seems to be a religious idea at the bottom of the recommendation to use the sputum lunæ (moon spittle) or sperm astrale (star semen), star mucus, in short of an efflux from the world of light above us, as first material for the work of our illumination.   [In many alchemistic recipes such things are recommended. Misunderstanding led to a so-called shooting star substance being eagerly hunted for.  What was found and thought to be star mucus was a gelatinous plant.]   So it is in this passage from John IX, 5, ff.: " As long as I am in the world I am the light of the world.   When he [Jesus] had thus spoken, he spat on the ground and made clay of the spittle and anointed the eyes of the blind man with the clay, and said unto him, Go, wash in the pool of Siloam [which is by interpretation: Sent].   He went his way, therefore, washed, and came seeing."   The transference of a virtue by the receiving of a secretion is a quite common primitive idea.

As Michael Maier (Symbola Aureae Mensae Lib.

XI) informs us, Melchior Cibinensis, a Hungarian priest, expressed the secrets of the forbidden art in the holy form of the Mass.   For as birth, life, exaltation, suffering in fire and then death were, as it were, ascribed to the Philosopher's Stone in black and gloomy colors, and finally resurrection and life in red and other beautiful colors, so he compared his preparation with the work of the salvation of man (and the " terrestrial " stone with the " celestial " stone), namely, with the birth, life, suffering, death and resurrection of Christ.   (Höhler, Herm. Phil., p. 156.)    The making of the Philosopher's Stone is, so to speak, the Imitation of Christ.

Hitchcock (H. A., p. 143) believes that Irenaeus Philaletha has clearly alluded in a passage of his writings to the two mental processes, analysis and synthesis, which lead to the same end.    " To seek the unity through Sol, I take it, is to employ the intellect upon the Idea of Unity, by analysis that terminates in the parts; whereas to study upon Mercury, here used for nature at large, is to work synthetically, and by combining the parts, reach an idea of the unity.   The two lead to the same thing, beginning as it were from opposite extremes; for the analysis of any one thing, completely made, must terminate in the parts, while the parts, upon a synthetical construction, must reproduce the unity.   One of the two ways indicated by Irenaeus is spoken of as a herculean labor, which I suppose to be the second, the reconstruction of a unity by a recombination of the parts, which in respect to nature is undoubtedly a herculean undertaking.   The more hopeful method is by meditation, etc.

Some of the writers tell us to put " one of the bodies into the alembic," that is to say, take the soul into the thought or study and apply the fire (of intellect) to it, until it " goes over " into spirit.   Then, " putting this by for use," put in " the other body," which is to be subjected to a similar trial until it " goes over " also; after which the two may be united, being found essentially or substantially the same.

The two methods of which Irenaeus speaks are also called in alchemy (with reference to chemical procedures) the wet and the dry ways.   The wet way is that which leads to unity through mental elaboration.   The philosophy of the Indian didactic poetry Bhagavad-Gita also knows the two ways and calls them Samkhya and Yoga.

" Thinking (Samkhya) and devotion (Yoga) separate only
     fools, but not the wise.
Whoever consecrates himself only to the One, gets both
     fruits.
Through thinking and through devotion the same point is
     reached,
Thinking and devotion are only One, who knows that, knows
     rightly."   [Bh-G. V. 4ff.

" Samkhya " and " Yoga " have later been elaborated into whole philosophical systems.   Originally, however, they are merely " different methods of arriving at the same end, namely the attainment of the Atman [all spirit] which on the one hand is

spread out as the whole infinite universe and on the
other is to be completely and wholly found in the
inner life.    In the first sense Atman can be gained by
meditation on the multiplex phenomena of the universe and their essential unity, and this meditation is
called Samkhya [from sam + khya, reflection, medi-
tation]; on the other hand, Atman is attainable by
retirement from the outer world and concentration
upon one's own inner world and this concentration is
called Yoga.    (Deussen, Allg. Gesch. d. Phil., I, 3,
p. 15.)

For the practice of alchemy a moral behavior is
required, which is hardly necessary as a precondition
of merely chemical work.    The disciple of the art is
to free his character, according to the directions of
the masters from all bad habits, especially to abjure
pride, is diligently to devote himself to prayer, perform works of love, etc.; no one is to direct his senses
to this study if he has not previously purified his
heart, renounced the love of worldly things, and sur-
rendered himself completely to God.    (Höhler,
Herm. Phil., pp. 62 ff.)

The sloppers, who strive to make gold in a chemical laboratory often waste in it their entire estate.
The adepts, however, assure us that even a poor man
can obtain the stone; many, indeed, say the poor
have a better materia than the rich.    Rom. II, 11:
"For there is no respect of persons with God."
Matth. xix, 24:  "It is easier for a camel to go
through the eye of a needle than for a rich man to
enter into the kingdom of God."    The alchemist

Khunrath says somewhere, the cost of making gold amounts to thirty dollars; we understand this when we remember that Jesus was sold for thirty pence.

Ruland (Lex., p. 26) defines alchemy very finely: [In reference to Tab. Smar., 9] " Alchemy is the separation of the impure from a purer substance." This is quite as true of the chemical as of the spiritual alchemy.

Why the hermetic philosophers write not literally but in figures may be accounted for in several ways. We should first of all remember that because of their free doctrine, which was indeed not at variance with true Christianity but with the narrow-minded church, they had to fear the persecution of the latter, and that for this reason they veiled their teachings. Hitchcock notices also a further point. The alchemists often declare that the knowledge of their secret is dangerous (for the generality of people). It appears that they did not deem that the time was ripe for a religion that was based more on ideal requirements, on moral freedom, than on fear of hell fire, expectation of rewards and on externally visible marks and pledges. Besides we shall see later that a really clear language is in the nature of things neither possible nor from an educational point of view to be recommended.

Still the mystical purpose of the authors of those times when the precautionary measures were not necessary appears clearer under the alchemistic clothing, although no general rule applying to it can be set forth. Other reasons, e.g., intellectual and conventional ones, influenced them to retain the symbolism.

Very clearly mystical are the writings of a number of hermetic artists, who are permeated by the spiritual doctrine of Jacob Boehme. This theosophist makes such full use of the alchemistic symbolism, that we find it wherever we open his writings. I will not even begin to quote him, but will only call the reader's attention to his brief and beautifully thoughtful description of the mystical process of moral perfection, which stands as " Processus " at the end of the 5th chapter of his book, " De Signatura Rerum." (Ausg., Gichtel Col., 2218 f.)

An anonymous author who has absorbed much of the " Philosophicus Teutonicus," wrote the book, " Amor Proximi," much valued by the amateurs of the high art. It does not require great penetration to recognize this pious manual, clothed throughout in alchemistic garments, as a mystical work. The same is true of the formerly famous " Wasserstein der Weisen " (1st ed. appeared 1619), and similar books. Here are some illustrative pages from " Amor Proximi ":

" This $\triangledown$ [$\triangledown$ of life] is now the creature not foreign or external but most intimate in every one, although hidden. . . . See Christ is not outside of us, but intimately within us, although hidden." (P. 32.)

" Whoever is to work out a thing practically must first have a fundamental knowledge of a thing; in order that man shall macrocosmically and magically work out the image of God, all God's kingdom, in himself; he must have its right knowledge in himself. . . ." (P. 29.)

" Christ is the great Universal; [The Grand Mastery is also called by the alchemists the " universal "; it tinctures all metals to gold and heals all diseases (universal medicine) ; there is a somewhat more circumscribed " particular," which tinctures only a special metal and cures only single diseases.] who says: ' Whoever will follow me and be my disciple (i.e., a particular or member of my body), let him take up his ✠ and follow me. Thus one sees that all who desire to be members of the great universal must each partake according to the measure of his suffering and development as small specific remedies." (Pp. 168 ff.)

" Paracelsus, the monarch of Arcana, says that the stars as well as the light of grace, nowhere work more willingly than in a fasting, pure, and free heart. As it is naturally true that the coarse sand and ashes cannot be illumined by the sun, so the SUN of righteousness cannot illumine the old Adam. It is then that the sand and ashes [the old Adam] are melted in the △ [of the ✠] again and again, that a pure glass [a newborn man] is made of it; so the ☉ can easily shoot its rays into and through it and therefore illumine it and reveal the wonder of its wisdom. So man must be recast in ✠ △ [cross-fire], so that the rays of both lights can penetrate him; otherwise no one will become a wise man." (P. 96 ff.)

Beautiful expositions of alchemy that readily make

manifest the mystical content are found also in the English theosophists Pordage and his followers, in particular Jane Leade (both 17th century). Their language is clearer and more lucid than Jacob Boehme's. Many passages appropriate to this topic might be here cited; but as I shall later take up Leade more fully, I quote only one passage from Pordage (Sophia, p. 23) :

" Accordingly and so that I should arrive at a fundamental and complete cleansing from all tares and earthiness . . . I gave over my will entirely to its [wisdom's] fiery smelting furnace as to a fire of purification, till all my vain and chaff-like desires and the tares of earthly lust had been burnt away as by fire, and all my iron, tin and dross had been entirely melted in this furnace, so that I appeared in spirit as a pure gold, and could see a new heaven and a new earth created and formed within me."

Out of all this, taken in conjunction with the following chapter, it will be evident and beyond question that our Parable must also be interpreted as a mystical introduction.

## ROSICRUCIANISM AND FREEMASONRY

THE previous chapter has shown that there was a higher alchemy — it was furthermore regarded as the true alchemy — which has the same relation to practical chemistry that freemasonry has to practical masonry. A prominent chemist who had entered into the history of chemistry and that of freemasonry once wrote to me: " Whoever desires to make a chemical preparation according to a hermetic recipe seems to me like a person who undertakes to build a house according to the ritual of Freemasonry."

The similarity is not a chance one. Both external and internal relations between alchemy and freemasonry are worthy of notice. The connection is partly through rosicrucianism. Since the Parable, which shall still be the center of our study, belongs to rosicrucian literature (and indeed is probably a later development of it), it is fitting here to examine who and what the Rosicrucians really were. We cannot, of course, go into a thorough discussion of this unusually complex subject. We shall mention only what is necessary to our purpose. I shall not, however, be partial, but treat of both the parties which are diametrically opposed in their views of the problems of rosicrucian history. It will be shown

173

that this disagreement fortunately has but small influence upon our problem and that therefore we are relieved of the difficult task of reaching a conclusion and of bringing historical proof for a decision which experienced specialists — of whom I am not one — have so signally failed to reach.

Rosicrucians are divided into those of three periods, the old, who are connected by the two chief writings, " Fama " and " Confessio," that appeared at the beginning of the 17th century; the middle, which apparently represents a degeneration of the original idealistic league, and finally, the gold crossers and rose crossers, who for a time during the 18th century developed greater power. The last Rosicrucians broke into freemasonry for a while (in the second half of the eighteenth century) in a manner almost catastrophic for continental masonry, yet I observe in anticipation that this kind of rosicrucian expansion is not immediately concerned with the question as to the original relation of freemasonry and rosicrucianism. We must know how to distinguish the excrescence from the real idea. Rosicrucianism died out at the beginning of the 19th century. The rosicrucian degrees that still exist in many systems of freemasonry (as Knight of the Red Cross, etc.) are historical relics. Those who now parade as rosicrucians are imposters or imposed on, or societies that have used rosicrucian names as a label.

Many serious scholars doubt that the old Rosicrucians ever existed as an organized fraternity. I refer to the article Rosenkreuz in the " Handbuch der Freimaurerei " (Lenning), where this skeptical view is dominant. Other authors, on the contrary, believe in the existence of the old order and think that the freemasons who appeared in their present form in 1717 are the rosicrucians persisting, but with changed name. Joh. Gottl. Buhle, a contemporary of Nicolai, had already assumed that the rosicrucian Michael Maier introduced rosicrucianism into England, and that freemasonry began then especially with the coöperation of the Englishman Robert Fludd (1574–1637). Ferdinand Katsch warmly defended the actual existence of the old rosicrucian fraternity with arguments, some of which are disputed. He names with certainty a number of people as " true rosicrucians," among them Julianus de Campis, Michael Maier, Robert Fludd, Frisius or Frizius, Comenius (Katch, p. 33). Rosicrucianism turned into freemasonry for practical reasons. As the most outstanding imposters represented themselves as rosicrucians this name was not conserved. The wrong was prevented, in that the true rosicrucians withdrew as such and assumed a different dress.

Generally we imagine a different origin of freemasonry. We are accustomed to look for its beginnings in practical masonry, whose lodges can be traced back to the fourteenth century. The old unions of house builders were joined by persons who were not actual workers but lay members, through whom spiritual power was added to the lodges. At the beginning of the eighteenth century the old work-

ing masonry was transformed into the spiritual symbolical freemasonry, but with a continuance of its
forms.   At that time in London the building lodges
had diminished to four.   These were united on June
24 (St. John's Day), 1717, and chose Anton Sayer
for their grand master.   That is the origin of Freemasonry as it exists to-day.

This derivation is and will be considered unsatis-
factory by many, however much it may satisfy the
merely documentary claims.   The attempt to make
it better required an inventive phantasy and this was
not always fortunate in its attempts.   The rosicrucian theory cannot be dismissed off hand, especially
if we conceive it in a somewhat broader sense.   In
agreement with Katsch, Höhler (Herm. Phil., p. 6)
recalls how generally people were occupied in the
16th and 17th centuries in the whole of western Eu-
rope with cabala, theosophy, magic (physics), astrology and alchemy, and indeed this held true of
higher and lower social strata, scholars and laymen,
ecclesiastic and secular.   " The entire learned theology turned on cabala.   Medicine was based on
theosophy and alchemy and the latter was supposed
to be derived from theosophy and astrology."
Höhler, in one respect, goes further than Katsch and
conjectures:   " Freemasonry had its roots in the
chemical societies of the 16th and 17th centuries, in
which all those things were fostered that constituted
the science of that day."   This theory is incompar-
ably more open to discussion than if one attempts to
confine the origin to the insecure base of rosicrucian-

ism. We shall learn to appreciate more fully the significance of the chemical societies.

In connection with the question, important for us, as to the position of the alchemy of the rosicrucians (whether they lived only in books or as an actual brotherhood), it is worth while to glance at the literature.

Joachin Frizius, whom some think identical with Fludd, writes in the " Summum Bonum, quod est verum Magiae, Cabalae, Alchymiae, verae Fratrum Roseae Crucis verorum subjectum " (first published in Frankfort, 1629):

"Aben ( אבן ) means a stone. In this one cabalistic stone we have the Father, Son and Holy Ghost . . . for in Hebrew Ab ( אב ) means Father and Ben ( בו ) Son. But where the Father and Son are present there the Holy Ghost must be also. . . . Let us now examine this Stone as the foundation of the macrocosm. . . . Therefore the patriarch Jacob spake, ' How dreadful is this place. This is none other but the house of God,' and rose up and took the stone that he had put for his pillow and poured oil upon the top of it, and said, ' This stone that I have set for a pillar shall be God's house, etc.' If therefore a God's house, then God is in that place or else his earthly substance. Here it was that the patriarch, as he slept on this stone, conserved something divine and miraculous, through the power of that spirit-filled stone which in its corporeality is similar to the relation of the body to the soul. But the spiritual stone was Christ; but Christ is the eternal wisdom, in which as the scripture says are many mansions, which are undoubtedly distinguished on account of the different grades of grace and blessedness. For blessedness follows wisdom or knowledge, the higher and more we know the farther we go towards the Godhead." (Summ. Bon., pp. 17 ff.)

" Thereupon it clearly appears who this macrocosmic Stone Aben . . . really is, and that his fiery spirit is the foundation stone of all and given for all (sit lapis seu petra catholica atque universalis) . . . which was laid in Zion as the true foundation, on which the prophets and the apostles as well have built, but which was also to the ignorant and wicked builders a stumbling block and bone of contention. This stone therefore is Christ who has become our Cornerstone. . . ." (Summ. Bon., p. 19.) " If we consider now the stone Aben in its significance for the microcosmos . . . we shall soon be sure that as a stone temple of God it can have no less value for every outer man in so far as the Holy Ghost also reserves a dwelling in him forever." (Summ. Bon., p. 20.)

" That is also the reason why the stone Aben appears in double form (quod ambae petrae), that is, in the macrocosmic and in the microcosmic. . . . For the spiritual stone is Christ that fulfills all. So we also are parts of the spiritual stone and such are also living stones, taken out of that universal stone (a petra illa catholica excisi). . . ." (Summ. Bon., p. 20.) Here again we have the alchemistic dis-

tinction between the universal and the particular, and
the like distinction is also expressed by the opposition of the celestial and the terrestrial stones.   The
second chapter of I Peter speaks of the living stone.
I Corinthians x, 4, says likewise: "And did all
drink of that spiritual Rock that followed them and
that Rock was Christ." Alchemistically expressed
it is called aurum potabile (drinkable gold).

"But," now you ask, "where then is all the gold
with which those alchemists [Fama] glitter so fa-
mously?" So we answer you. . . . "Our gold is
indeed not in any way the gold of the multitude, but
it is the living gold, the gold of God. . . . It is wisdom, which the psalmist means, Ps. xii, 6, ' The
words of the Lord are pure words as silver tried in
a furnace of earth, purified seven times.' If you
now wish . . . to put before yourself the true and
actual animal stone, then seek the cornerstone, which
is the means of all change and transformation, in
yourself." (Summ. Bon., pp. 34 ff.)

"Finally the brother works towards the consummation of his labors in the form of a master builder
(*denique sub architecti figura operatur frater ad
huius operis perfectionem*). . . . Only for the better carrying out of our building and thereby to attain
the rose-red bloom of our cross concealed in the center of our foundation . . . we must not take the
work superficially, but must dig to the center of the
earth, knock and seek." (Summ. Bon., p. 48;
Trans. Katsch, pp. 413 ff.) Just after that he
speaks of the three dimensions, height, depth, and

breadth.   The masonic symbolism is accompanied clearly enough in the " Summum Bonum " by the alchemistic.   Notice the knocking and seeking, and what is mentioned in the doctrines about the form of the Lodge.   Immediately thereafter is a prolix discussion of the geometric cube.

Frizius and Fludd contribute also a letter supposed to have been sent by rosicrucians to a German candidate.   It says, " Since you are such a stone as you desire, and such a work . . . cleanse yourself with tears, sublimate yourself with manners and virtues, decorate and color yourself with the sacramental grace, make your soul sublime toward the subtile meditation of heavenly things, and conform yourself to angelic spirits so that you may vivify your moldering body, your vile ashes, and whiten them, and incorruptibly and painlessly gain resurrection through J[esus] C[hrist] O[ur] L[ord]."   In another passage:   " Be ye transformed, therefore, be ye transmuted from mortal to living philosophic stones."

In the " Clavis Philosophiae et Alchymiae Fluddanae " (published in Latin in 1633), are passages like the following:   " Indeed every pious and righteous man is a spiritual alchemist. . . . We understand by that a man who understands not only how to distinguish but with the fire of the divine spirit to separate [spagiric art] the false from the true, vice from virtue, dark from light, the uncleanness of vice from the purity of the spirit emulating God.   For only in this way is unclean lead turned into gold."   (P. 75.)   " If one now ventures to say that the

Word of Christ or the Holy Ghost of wisdom dwells in the microcosmic heaven [i.e., in the soul of man] we should not decry the blind children of the world as godless and abandoned. [But certainly the divine spirit is, as is later averred, the rectangular stone in us, on which we are to build.] This divine spark is, however, continuous and eternal; it is our gold purchasable of Christ. . . . So it happens in accordance with the teachings of Christ, or the Word become flesh, that if the true alchemists keep on seeking and knocking, they attain to the knowledge of the living fire." (P. 81.) So again the important knocking and seeking of masonic symbolism, and this indeed, for the purpose of learning to know a fire.

In reference to the really elevating thoughts of the " Summum Bonum," Katsch, enthusiastic about these ideas, exclaims: "What language, what an unflinching courage, what a dignified humility. Even the most reluctant will not be able to avoid the admission that here quite unexpectedly he has . . . met the original and ideal form of freemasonry."

The comparison of masonry and alchemy remains true even if we work more critically than Katsch, who is accused of many inaccuracies. I recall for instance the later researches of the thorough and far-seeing Dr. Ludwig Keller.

For the illumination of the darkness that has spread over the past of freemasonry, Keller shows us (B. W. and Z., pp. 1, 2) the rich material of symbolism that is offered the diligent student, first of all in the very copious literature, printed matter, and especially in the manuscripts, that is known by the name of Chemistry or Alchemy.

In the symbols of the alchemists, the rosicrucians, the Lodges, etc., "we meet a language that has found acceptance among all occidental peoples in analogous form, not indeed a letter or word language, but a language nevertheless, a token or a symbol language of developed form, which is evident even in the rock temples of the so-called catacombs, once called latomies and loggie. The single images and symbols have something to say only to the person who understands this language. To the man who does not understand it, they say nothing and are not expected to say anything."

In reference to the symbol and image language, which was comprehensible only to the initiated, we think naturally of the ancient mysteries. The religious societies of the oldest Christians, in the centuries when Christianity belonged in the Roman Empire to the forbidden cults, found a possibility of existence before the law in the form of licensed societies, i.e., as guilds, burial unions, and corporations of all sorts. The primitive Christians were not the only forbidden sects that sought and found this recourse. Under the disguise of schools, trade unions, literary societies, and academies, there existed in the jurisdiction of the Roman Empire, and later inside of the world church, organizations that before the law were secular societies, but in the minds of the initiated were associations of a religious character. Within these associations there appeared very early

a well developed system of symbols, which were
adopted for the purpose of actually maintaining,
through the concealment necessitated by circumstances, their unions and their implements and customs — symbols that they chose as cloaks and that
in the circle of the initiated were explained and interpreted according to the teachings of their cult.

Valuable monuments of this symbolism are preserved in the vast rock temples that are found in
Egypt, Syria, Asia Minor, Sicily, and the Apennine
peninsula, in Greece, France, and on the Rhine, and
these vaults, which in part also served the early
Christians as places of worship, show in their images
and records and in their architectural form so close
a resemblance that they must be acknowledged as
the characteristic of a great religious cult extending
over many lands, which has had consistent traditions
for the use of such symbols and for the production of
these structures.

Many of these symbols, it should be noted in passing, are borrowed from those tokens and implements
of the building corporations, which were necessary
to the completion of their buildings (Keller, l. c.,
p. 4). An important part was played even in the
early Christian symbolism by the sacred numbers
and the figures corresponding to them, a group of
educational symbols which we find likewise in the
pythagorean and platonic schools. It is known that
the symbolical language of the subterranean rock
temples, some of which were used by the earliest
Christians for their religious worship, are closely

connected with the pythagorean and platonic doctrines.    From the year 325 A. D. on, every departure
from the beliefs of the state church was considered
a state offense.    So those Christians who retained
connection with the ancient philosophic schools were
persecuted.    In the religious symbol language of the
church, the sacred numbers naturally began to disappear from that time.    In the writings of Augustine begins the war on the symbolic language, whose
use he declared a characteristic of the gnostics.    In
spite of the suppression the doctrines of the sacred
numbers continued through all the centuries in religious use, in quiet but strong currents which flowed
beside the state church.    The sect names, which
were invented by polemic theology for the purpose of characterizing methods that were regarded
as imitations of the gnostics, are of the most varied
kinds; it may be enough to remember that in all those
spiritual currents, that like the old German mysticism, the earlier humanism, the so-called natural philosophy, etc., show a strong influence of platonic
thinking, the doctrines of the sacred numbers recur,
in a more or less disguised form, but yet clearly recognizable.    (Keller, Heil. Zahl., p. 2.)

As the old number symbolism constitutes a part
of the hieroglyphics of alchemy, I shall pause a moment to consider them.    The use of mathematical
and geometrical symbols proceeds from the use of
the simplest forms, points and lines, but in all cases
where the object is not a representation in the flat but
in space, both the points and lines are replaced by

plastic forms, i.e., forms of cylinders, spheres, bars, rings, cubes, etc. From this point it was but a short step to the use of trees, leaves, flowers, implements, and other things that showed similarities in form. Pillars are specially noticeable for the symbolism of the ceremonial chamber. In all cases where points and lines occur in images and drawings, pillars are found in the plastic representation of thoughts and symbols. They form the chief element of the organization of cults in academies and museums, and justify the names of colonnade, stoa, portico, and loggia, which occur everywhere; besides the special designation like Οἶκυς αἰοήιος, etc.

For symbolism, too, which served as the characterization of the forms of organization and the building up of the fraternity into degrees, lines were useless, but in place of lines and points are found plastic forms which were at their disposal in carpenters' squares, crossed bars, etc. (Keller, l. c., p. 10.)

As the circle symbolized the all and the eternal or the celestial unity of the all, and the divinity, so the number one, the single line, the staff or the scepter, represented the terrestrial copy of the power, the ruling, guiding, sustaining and protecting force of the personality that had attained freedom on earth.

The sun or gold symbol ⊙ corresponds in alchemy to the divine circle and the same circle occurs in other symbols of the art, as in ♀ ☿, etc.

Duality, the Dyas, represents in contrast to the celestial being the divided terrestrial being that is dominated by the antagonism of things and is only a transitory, imperfect existence; the opposites, fluid and solid, sulphur and mercury, dry and wet, etc.

In the symbol of the trinity, which frequently occurs in the form of a triangle (three points united by three straight lines), is shown how the divided and sensuous nature is led by the higher power of the number 3 to a harmony of powers and to a new unity. The symbol of reason attaining victory over matter becomes visible. A representation of trinity is possible by means of the conventional cross. We can see in it two elements of lines which by their unification or penetration give the third as the point of intersection. More generally the cross is conceived as quinity (fiveness) — i.e., 4+1ness (in alchemy four elements which are collected about the quinta esentia). A cross in which unity splits into duality so that trinity results, is Y, which is called the forked cross. From unity grows duality, that is, nature divides into spirit and matter, into active and passive, necessity and freedom. The divided returns through trinity to unity. In alchemy we have the symbol REBIS, the hermaphrodite with the two heads. The ancient symbol was later conceived, by purposive concealment or by more accidental interpretation, as the letter Y, just as the symbol of the three lines △ or △ and the like gradually appears to have become an A, as it is found frequently in the catacombs. Keller refers (l. c., p. 14) also especially to the reduplication of the carpenter's square, which is found likewise in the old Latomies (Gk. ⚌

quarries) and has the appearance of two intersecting opened circles. I do not need to call attention to the masonic analogue; in alchemy we have here the interpenetration of △ and ▽, i.e., ⊗, which is among others the symbol for the material of the stone. [△ and ▽ are the symbols for the elements fire and water. Fire and water, however, mean also the famous two opposites, that are symbolized quite as well by warm and cold, red and white, soul and body, sun and moon, man and woman.] With regard to the six points, in alchemy ⊗ is also called chaos in contrast to ⊗, which denotes cosmos, just as alum ○ on account of its lack of a center (God, belief, union), is incomplete beside ☉. In the catacombs the triangle is found also in multiple [five fold] combination, ⊗ .

Four lines, somewhat in the form of a rectangle, define the limited space of the terrestrial world with the accessory meaning of the holy precinct, house, temple. In masonry, □ is well known as the lodge. The rectangle is related to the cube. I mention therefore in this place the cubic stone, the mighty masonic symbol, whose equivalent in alchemy will be discussed.

By a commonly used change of significance the number 5 is symbolized by 5-leaved plants (rose, lily, vine). " The flowers, however, and the garden in which they grow, early served as symbols of the Fields of the Blessed or the ' better country ' in which dwell the souls passing through death to life; in antithesis to the terrestrial house of God, the

temple built with hands, which was represented by
the rectangle ☐, the holy number 5 denoted the
celestial abodes of the souls that had attained perfection, and therefore represented both the House
of Eternity or the City of God and the Heavenly
Jerusalem.   The holy pentagram in the form of the
rose, not only in the ancient but in the early Christian world, decorated the graves of the dead, that
in their turn symbolized the gardens of the blessed.
And the significance that the academies and loggia
attributed to the pentagram placed in the rose is explained by the fact that their religious festival was
closely connected with this emblem.   Already in the
ancient world at the festival of St. John, the rose
feast or rhodismus or Rosalia was celebrated, at
which the participants adorned themselves with roses
and held religious feasts."   (Keller, l. c., p. 21.)

As already mentioned, the cross, i.e., the Greek
cross with its four equal arms, expresses the number
five.   It is interesting that already in the ancient
number symbolism, rose and cross appear united, a
fact which I mention here in view of the later connection of these two objects.

The semicircle or moon is an emblem of borrowed
light.   Besides the circles or spheres, the symbols
of eons (divine beings, powers) that are enthroned
in the ether as eternal beings, the human soul — the
psyche or anima, which does not coincide with reason
or the purified soul — appears as a broken circle.
As the sun and its symbol, the ragged circle, sym-
bolize the eternal light, the half circle is, as it were,

the symbol of that spark of light that slumbers in
the soul of man, or, as the alchemists often say, the
hidden fire that is to be awakened by the process.
If we reflect that in this symbolism the cross expresses a penetration, the alchemic symbol ☿ is explained. It is now quite interesting that the like
connection appears in the subterranean places of worship in this form ♀ (l. c., p. 27). Keller calls it a
symbol of the all and the soul of man.

The number 7 (seven planets, etc.) also is of some
importance in the old latomies. It is noteworthy
besides that sun and moon usually appear as human
forms; the sun wears on its head a crown or garland
or beaming star, while the moon image is wont to
carry the symbol ☽ . Alchemy, too, likes to represent ☉ and ☽ as human, and indeed frequently as
crowned figures, sometimes as a royal bridal couple.

The ancient lore of the sacred numbers breathes a
spirit that may be embodied in the following words:
The soul of man, which through resignation or meek-
ness, as they used to say then, is impelled onward
to purity and union with the Eternal, has in itself
a higher life, which cannot be annihilated by death.
The doctrine of the infinite value of the soul . . .
and of God's entering into the pure soul of man
forms the central point of the thought of religious
fellowship. Neither for sacrifice, which the state
religions practice, nor for the beliefs in demons, by
which the masses are controlled, nor for the idea
of priesthood as means of salvation, was there a
place in this system, and not a trace of such a belief

is demonstrable in this religion of wisdom and virtue.   (l. c., p. 33.)

Besides the early Christian ideal, which recognized and encouraged the connection between the
teachings of Christ and the ancient wisdom of pla-
tonism, there was in early times another which emphasized and endeavored to develop the antithesis
more than the connection.   From the time when the
new Christian state church came to life, and sacri-
ficial religion and the belief in devils and the priesthood were restored, a struggle of life and death developed between the church and the so-called philosophic schools.   " The fraternity saw that it had to
draw down the mask still further over its face than
formerly, and the ' House of the Eternal,' the
' Basilika,' the ' Academies,' and the ' Museums ' became workshops of stone cutters, latomies, and loggia or innocent guilds, unions, and companies of
every variety.   But all later greater religious move-
ments and tendencies which maintained the old beliefs, whether they appeared under the names of
mysticism, alchemy, natural philosophy, humanism,
or special names and disguises, as workshops or societies, have preserved more or less truly the doctrine of the " sacred numbers " and the number symbolism, and found the keys of wisdom and knowledge in the rightly understood doctrine of the eternal harmony of the spheres."   (Keller, l. c., p. 38.)

Keller derives modern freemasonry from the academies of the renaissance, which, as we have just
heard, continued the spirit of the ancient academies.

Now it is interesting that the later branches of these religious societies (after the renaissance) took among others the form of alchemy companies and further that such fraternities or companies [as are not called alchemical], still employed symbols that we recognize as derived from alchemy. The hieroglyphics of alchemy appear to be peculiarly appropriate to the religious and philosophic ideas to be treated of. Rosicrucianism was, however, one of the forms into which alchemy was organized. It is further important that in just those societies of the beginning of the seventeenth century which outsiders called " alchymists " or " rosicrucians," the characteristic emblems of the old lodge appeared, as, for instance, the circle, the cubic stone, the level, the man facing the right, the sphere, the oblong rectangle (symbol of the Lodge), etc. (Keller, Zur Gesch. d. Bauh., p. 17.) These " alchymists " honored St. John in the same way as can be shown for the companies of the fifteenth century. I need not mention that modern masonry, in its most important form, bears the name of Masons of St. John.

From the beginning of the 17th century attempts were made inside the fraternity, as the company societies working in the same spirit may be called, to bring to more general recognition a suitable name for this company, which could also form a uniting bond for the scattered single organizations. The leaders knew and occasionally said that a respected name for the common interest would be advantageous. This view appears especially in the letters of

Comenius. It was then indeed an undecided question what nation should place itself at the head of the great undertaking. (Keller, in the M. H. der C. G., 1895, p. 156.) "As a matter of fact precisely in the years when in Germany the brothers had won the support of powerful princes and the movement received a great impetus, very decided efforts were made both to create larger unions and to adopt a unifying name. The founding of the Society of the Palmtree [1617] was the result of the earlier effort and the writings of Andreaes on the alleged origin and aims of the rosicrucians are connected with the other need. The battle of the White Mountain and the unfortunate consequences that followed killed both attempts, as it were, in the germ." (Z. Gesch, d. Bauh., p. 20.) Note by the way that the name of the " Fraternity of the Red Cross " was taken from symbols which were already employed in the societies. In regard to this it is quite mistaken accuracy to maintain that it was correctly called " Bruderschaft des Rosenkreutz " and not " des Rosenkreutzes," as the " Handbuch d. Freimaurerei," p. 259, emends it. Vatter Christian Rosenkreutz is indeed evidently only a composite legendary personage as the bearer of a definite symbolism (Christ, rose, cross), (and may have been devised merely in jest). The name does not come from the personality of the founder but the personality of the founder comes from the name. The symbols and expressions that lie at the foundations are the earlier.

The attempt mentioned, to find a common name, did not permanently succeed. The visionaries and " heretics " decried as " Rosicrucians " and " alchymists " were considered as enemies and persecuted. It is irrelevant whether there was an organized fraternity of rosicrucians; it was enough to be known as a rosicrucian. (Keller, Z. Gesch. d. B., p. 21.) The great organization did not take place until a great European power spread over it its protecting hand, i.e., in 1717, when in England the new English system of " Grand Lodges of Free and Accepted Masons " arose. (Keller, D. Soc. d. Hum., p. 18.) We see that Keller arrives by another and surer way than Katsch at the same result, and shows the continuity of the alchemists or rosicrucians and the later freemasons, if not in exactly the same way that Katsch has outlined it. In particular Keller gets along without the unproved statement that there were organized rosicrucians (outside of the later gold- and rose-crosses). He shows what is much more important, namely that there were societies that might have borne the name of rosicrucians (or any similar name).

Several interesting peculiarities should not be omitted, as for instance, that Leibniz, about 1667, was secretary of an alchemist's society (of so-called rosicrucians) in Nuremberg. Leibniz describes alchemy as an " introduction to mystic theology " and identifies the concepts of " Arcana Naturae " and " Chymica." (M. H. der C. G., 1903, p. 149; 1909, p. 169 ff.) In the laws of the grand lodge

" Indissolubilis " (17th and 18th centuries) there
are found as doctrinal symbols of the three grades,
the alchemistic symbols of salt (rectification, clarification), of quicksilver (illumination), and of sulphur (unification, tincture), used in a way that corresponds to the stages of realization of the " Great
Work." The M. H. d. C. G., 1909, p. 173 ff.
remarks that we should probably regard it only as
an accident, if there are not found, in the famous
hermetic chemical writings, similar signs with additions as would for experts, exclude all doubt as to
their purport.    In 1660 appeared at Paris an edition of a writing very celebrated among the followers of the art, " Twelve Keys of Philosophy," which
was ostensibly written by one Brother Basilius Valentinus.    In this edition we see at the beginning a remarkable plate, whose relation to masonic symbolism
is unmistakable (Figure 1).    In addition to the lowest symbol of salt (represented as cubic stone) there
is a significant reference to the earth and the earthly.
[I should note that besides $\ominus$ alchemy used $\square$ for
salt, in which there is a special reference to the
earthly nature of salt.    In Plato the smallest parti-
cles of the earth are cubical.    Salt and earth alternate in the terminology, just as mercury $\female$ and air $\triangle$
or water $\triangledown$ do; as sulphur $\spadesuit$ and fire $\triangle$; only, however, where it is permitted by the context.]    The
Rectification of the subject (man) taken up by the
Art, is achieved through the purification of the
earthly elements according to the indication of the
alchemists who call the beginning of the work " Vit-

Fig. 1—Cut above.

Fig. 2—Cut at right.

Fig. 3—Cut below.

riol," and form an acrostic from the initial letters
of this word: "Visita Interiora Terrae, Rectifi-
cando Invenies Occultum Lapidem" [= Visit the
interior of the earth; by purifying you will find the
hidden stone]. Half way up there floats the ☿
that has the value of a "union symbol" in the broth-
erhoods (as such, a symbol of fellowship) and left
and right of it is found the moon and sun or the
flaming star. Above is placed a triangle, in which
is a phœnix rising from the flames; and on the triangle stands the crowned Saturn or Hermes (in
masonry Hiram). On the left and right of this
kingly form, on whose breast and stomach are placed
planet symbols, we notice water in the shape of
drops (tears) and flames that signify suffering and
resurrection. "When we notice that not only the
principles of the old 'amateurs of the art' correspond with those of the 'royal art' [freemasonry],
but that the symbolism also is the same in all parts,
we recognize that the later masonic societies are only
a modern reshaping of the societies which dropped
the depreciated names of the alchemists in order to
appear in a new dress (l. c., p. 175). That the as-
sertion of the complete similarity of the symbolism
is not mere fancy, the following considerations (and
not those only in this section), will satisfactorily
demonstrate. In the following examples the words
showing it most clearly are italicized.

Alchemy was regarded by its disciples as a *royal
art*. Old sources show that the art of making gold
was revealed in Egypt only to the crown princes.

Generally only the kings' sons were informed by the priests concerning the magic sciences. The hermetics derived their art expressly from kings, Hermes, Geber, and the patriarchs of alchemy were represented as kings.

According to Khunrath (Amphitheatrum) prayer, work and perseverance lead to eternal wisdom by the mystical ladder of the *seven* theosophical *steps*. Perfect wisdom consists in the knowledge of God and his Son, in the understanding of the holy scriptures, in self knowledge and in knowledge of the great world and its Son, the Magnesia of the philosophers or the Philosopher's Stone. The mystical steps in general contain *three* activities, hearing (audire), persevering (perseverare), knowing (nosse et scire), that applies to *five* objects, so that we can distinguish *seven* steps in all. Only the pure may enter the temple of wisdom, only the *worthy* are intrusted with the secrets, the *profane,* however, must stay away.

In the fifth table of Khunrath's Amphitheatrum is pictured the seven pillared citadel of Pallas (Prov. IX, 1). At the entrance is a table with the legend Opera bona (= good works). Behind sits a man with the staff of Mercury. On each side is a *four sided pyramid,* on the top of the left one is the *sun,* on the right the *moon.* On the former stands the word *Fides* (= faith), on the latter *Taciturnitas* (= silence). Behind the man we read the word *Mysterion,* over the inner entrance *Non omnibus* (= not for all).

Alchemy frequently mentions two or three *lights*.
By these it understood ⊙ and ☽, ♎ ☿ ☉, light of
grace and light of nature, etc. The juxtaposition
of ⊙ ☽ and ☒ is interesting; no one can attain the
desired end before, through the *circular wheel* of
the elements, the fatness or the blood of the *sun,*
and the dew of the *moon* are by the action of *art*
and *nature,* united in one body in the image of the
*hexagram;* and this can take place only by the will
of the *Most High,* who alone imparts the unique
boon of the *Holy Ghost* and *priceless treasure* according to his especial mercy. The above mentioned
circular wheel is identical with the serpent that bites
its own tail; it is a power that always consumes and
always renews itself. This circle appears not to be
lacking in the flaming star; it is the round eye or the
likewise round fashioned " G," which latter looks
quite similar to the snake hieroglyph. The reference to Genesis has a good reason. Moreover, the
hexagram represents in cabbalistic sense the mystical
union of the male with the female potence △ with ▽ .
According to a rabbinical belief a picture is supposed
to be placed in the ark of the covenant alongside of
the tables of the laws, which shows a man and a
woman in intimate embrace, in the form of a hexagram. In cabbalistic writings, as for instance, in
those of H. C. Agrippa, we find the human form in
a star, generally inscribed in the pentagram. The
genitals fall exactly in the middle part and are often
made prominent by an added ☿ as male-female or
androgyne procreative power. One of the snake

shaped Egyptian hieroglyphs frequently turns into an Arabic ‏ج‎ , i.e., gimel. I do not know whether this fact has any significance here. With respect to the above passages that mention the " will of the Most High," I refer to the dialogue which concerns the " G "; e.g., " Does it mean nothing else? " " Something that is greater than you." " Who is greater than I? " etc. " It is Gott, whom the English call God. Consider this mysterious star; it is the symbol of the Spirit. . . . The image of the holy fire, etc."

REBIS is represented as an hermetic hermaphrodite. The already mentioned figure with the two heads (figure 2) is found (as Höhler relates) in a book that appeared in Frankfort in 1618, called " Joannes Danielis Mylii Tractatus III, seu Basilica Philosophica," though it is to be seen also in other books on alchemy. The hermaphrodite stands on a dragon that lies on a globe. In the right hand he holds a *pair of compasses,* in the left a *square.* On the globe we see a *square* and a *triangle.* Around the figure are the signs of the seven planets, with ☿ at the top. In a cut in the Discursus Nobilis of John of Munster we see *sun* and *moon,* at the middle of the top the *star* ✸ , also denoted by $Y =$ "γλζ ($=$ matter) surrounded by *rays.* (Höhler, Herm. Phil., p. 105.)

In the cabala, which has found admission into the idea of the alchemists and rosicrucians, no small part is played by *three pillars* and *two pillars.*

Tubal Cain was renowned as a great alchemist.

He was the patriarch of wisdom, a master of all kinds of brass and iron work. (Genesis IV, 22.) He had the knowledge not only of ordinary chemistry and of the fire required for it, but also of the higher chemistry and of the hidden elemental fire. After the flood there was no other man who knew the art but the righteous Noah, whom some call Hermogenes or Hermes, who possessed the knowledge of celestial and terrestrial things.

One devoted to art must be a *free man* (Höhler, l. c., p. 66). The *ordinale* of Norton establishes it more or less as follows: "The kings in the olden time have ordained that no one should learn the liberal sciences except the free and those of noble spirit, and any one who is devoted to them should devote his life most freely. Accordingly the ancients have called them the seven liberal arts, for whoever desires to learn thoroughly and well must enjoy a certain freedom."

Very frequently one finds in the alchemists images of *death:* grave, coffin, skeleton, etc. Thus in Michael Maier's, Atalanta Fugiens, the Emblema XLIV shows how the *king* lies with his crown in the *coffin* which is just *opened.* On the right stands a man with a turban, on the left two who open the coffin and let his joyful countenance be seen. In the Practica of Basilius Valentinus the illustration of the fourth key shows a coffin, on which stands a skeleton, the illustration of the eighth key (see Fig. 3), a grave from which half emerges a man with upright body and raised hands. [This reproduction and figure I owe to the kindness of Dr. Ludwig Keller and the publications of the Comenius Society.]   Two men are shooting at the well known mark, ☉, here represented as a target (a symbol much used in the old lodges), while a third is sowing.   (Parable of the sower and the seeds.)   The sign is a clever adaptation of the sulphur hieroglyph and is identical with the registry mark of the third degree of the Grand Lodge Indissolubilis.   The mark ∩ on the wall is also a symbol of the academy; it is the half circle, man, to whom the light is imparted and means, when occurring collectively, the fraternity.   The evident idea is of representing the exclusive society as enclosing wall.   The angel with the trumpet is the angel of the judgment day who awakes the dead.   With respect to the birds I refer to Matthew XIII, 4:   " And when he sowed, some seeds fell by the wayside and the fowls came and devoured them up."   In the text of Basilius Valentinus, the fourth key, there is mention of the rotting and falling to pieces with which we are familiar. The idea of dismemberment is not infrequently clearly expressed, more clearly than in our parable. Already in the oldest alchemistic manuals one operation is called the grave of Osiris.   One of the manuscripts cited by Berthelot (Orig., p. 60) says: " The dragon is the *guardian of the temple,* sacrifice him, *flay* him, cut his *flesh* from his *bones* and thou wilt find what thou seekest."   The dragon is also called Osiris, with whose *son* Horus-Harpocrates, the skillful Hermes, is also identified.   (Do we need

reference to requirements in the 3d degree? J. . . .
left his skin; . . . . B . . . . left his flesh . . . .;
M . . . . B. . . ., he lives in the Son.)

Here more clearly than anywhere else we see the
masonic symbolism combined with the myth of the
first parents or creation myth.  No matter where
it acts, the myth-making power never seems willing
to belie its laws.  Also the tree growing out of the
grave or the body of the dead ancestor is not want-
ing.  (". . . at the graves of our fathers."  " I
was accused of a terrible crime.")  It is the acacia
whose presence is rationalized apparently for the
purpose of forming a sign by which to find again the
place of the hastily buried.

An Egyptian fable tells of two brothers.  The
younger, Bata, was *falsely accused* by his sister-in-
law (as was Joseph by Potiphar's wife).  His
brother Inpw (Anepu) consequently pursued him.
The sun god made a mighty flood that separated the
pursuer from the pursued.  Bata castrated himself
and threw his organ of generation into the water,
where it was swallowed by a fish.  Bata's heart later
in the story is changed into a blossom of an acacia
or a cedar.  [I naturally lay no stress on the accident that the acacia occurs here.  The point is that
the tree is a symbol of life.]  Bata is reconciled
with Inpw and at parting relates to him that a mug
of beer is to serve as a symbol of how the brother
fares, who is dwelling afar off.  If the beer foams
he is in danger.  Bata's wife has the acacia tree, on
which Bata's heart is a blossom, felled, and as a

result Bata dies.    By means of the mug Inpw learns
of Bata's peril and departs to look for his younger
brother.    Inpw *finds the fallen acacia* and on it a
berry that is the heart of his brother transformed.
Bata *comes to life again* and transforms himself
into an ox.    His wife has the ox butchered on the
pretext of wishing to eat its liver.    Two drops of
blood fall from the cut throat of the ox upon the
ground and are changed into two peach trees.
Bata's wife has the two peach trees felled.    A chip
flies into her mouth.    She swallows it and becomes
pregnant by it.    The child that she bears is the
reincarnated Bata.    He therefore *lives again in his
son as the child of a widow.*

The second fragment of the Physica et Mystica
of Pseudo Democritus, that Berthelot cites (Orig.,
p. 151) relates that the *master died without having
initiated Democritus into the secrets of knowledge.*
Democritus conjured him up out of the underworld.
The spirit cried:    " So that is the reward I get for
what I have done for thee."    To the questions of
Democritus he answered, " The books are in the
temple."    They were not found.    Some time thereafter, on the occasion of a festival, they saw a
*column* crack open, and in the opening they found
the books of the master, which contained three mystic axioms:    " Nature pleases herself in Nature;
Nature triumphs over Nature; Nature governs Nature."

The quotations show, to be sure, only superficially
the interrelation of alchemy and freemasonry.    The

actual affinity lying behind the symbolism, which, moreover, our examination of the hermetic art has already foreshadowed, will be treated later.

We could also posit a psychological interrelation in the form of an " etiological assumption " according to the terminology of psychoanalysis. It would explain the temporary fusion of alchemistic rosicrucianism with freemasonry. The rosicrucian frenzy would never have occurred — so much I will say — in masonry, if there had been no trend that way. Some emotional cause must have existed for the phenomenon, and as the specter of rosicrucianism stalked especially on the masonic stage, and indeed was dangerous to it alone, this etiological assumption must be such as to furnish an effective factor in masonry itself, only in more discreet and wholesome form. In masonry psychological elements have played a part which if improperly managed might degenerate, as indeed they did when gold- and rose-crossism was grafted on masonry. It appears to me too superficial to explain the movement merely from the external connection of rosicrucianism and the masonic system. Although the observation is quite just, it does not touch the kernel of the matter, the impulse, which only psychology can lay bare. Freemasonry must have felt some affinity with rosicrucianism, something related at the psychical basis of the mode of expression (symbolism, ritual) of both. Only the modes of expression of rosicrucianism are evidently more far reaching or more dangerous in the sense that they (the leadership of loose companions

always presupposed) could sooner incite weaker
characters to a perverted idea and practice of it.

That rosicrucianism in its better aspect is identical
with the higher alchemy, can no longer be doubted
by any one after the material here offered.    The
common psychological element is shown when, as
will be done in later parts of this book, we go into
the deeper common basis of alchemy and freemasonry.    Then first will the sought-for " etiological
assumption " attain to its desired clearness.    But
already this much may be clear: that we have in
both domains, structures with a religious content,
even though from time to time names are used which
will veil these facts.    I add now in anticipation a
statement whose clear summing up has been reserved
for psychoanalysis, namely that the object of religious worship is regularly to be regarded as a symbol of the libido, that psychologic goddess who rules
the desires of mankind — and whose prime minister
is Eros.    [Libido is desire or the tendency toward
desire, as it controls our impulsive life.    In medical language used mainly for sexual desire, the concept of libido is extended in psychoanalysis (namely
by C. G. Jung) to the impelling power of psychic
phenomena in general.    Libido would therefore be
the inner view of what must in objective description
be called " psychic energy."    How it could be given
this extension of meaning is seen when we know the
possibilities of its transformation and sublimation,
a matter which will be treated later.]    Now if the
libido symbol raised up for an ideal is placed too

nakedly before the seeker, the danger of misunderstanding and perversion is always present. For he
is misled by his instincts to take the symbol verbally,
that is, in its original, baser sense and to act accordingly. So all religions are degenerate in which one
chooses as a libido symbol the unconcealed sexual
act, and therefore also a religion must degenerate,
in which gold, this object of inordinate desire, is
used as a symbol.

What impels the seeker, that is, the man who
actually deserves the name, in masonry and in alchemy, is clearly manifested as a certain dissatisfaction. The seeker is not satisfied with what he
actually learns in the degrees, he expects more, wants
to have more exhaustive information, wants to know
when the " real " will be finally shown. Complaint
is made, for example, of the narrowness of the meaning of the degrees of fellowship. Much more important than the objective meaning of any degree
is the subjective wealth of the thing to be promoted.
The less this is, the less will he " find " even in the
degrees, and the less satisfied will he be, in case he
succeeds in attaining anything at all. To act here
in a compensating way is naturally the task of the
persons that induce him. But it is the before mentioned dissatisfaction, too, which causes one to expect wonderful arts from the superiors of the higher
degrees; an expectation that gives a fine opportunity
for exploitation by swindlers who, of course, have
not been lacking in the province of alchemy, exactly
as later at a more critical time, in the high degree

masonry. Who can exactly determine how great a
part may have been played by avarice, ambition,
vanity, curiosity, and finally by a not unpraiseworthy
emotional hunger?

The speculators who fished in the muddy waters
of late rosicrucianism put many desirable things as
bait on the hook; as power over the world of spirits,
penetration into the most recondite parts of nature's
teachings, honor, riches, health, longevity. In one
was aroused the hope of one of these aims, in another
of another. The belief in gold making was, as already mentioned, still alive at that period. But it
was not only the continuance of this conviction that
caused belief in the alchemistic secrets of the high
degrees, but, as for instance, B. Kopp shows (Alch.
II, p. 13) it was a certain metaphysical need of the
time.

It will have been noticed that with all recognition
of its abuses I grant to rosicrucianism, as it deserves,
even its later forms, an ideal side. To deny it were
to falsify its true likeness. Only the important difference must be noted between an idea and its advocates alchemy and the alchemists, rosicrucianism and
the rosicrucians. There are worthy and unworthy
advocates; among the alchemists they are called the
adepts or masters and the sloppers and sloppy workers. Since in our research we are concerned with
the hermetic science itself, not merely with the mis-
directions undertaken in its name, we should not let
ourselves be involved in these. And as for us the

spiritual result (alchemy, rosicrucian thoughts, masonic symbolism, etc.) is primarily to be regarded and not the single persons advocating it, the question is idle as to whether the earliest rosicrucians had an organized union or not. It is enough that the rosicrucians are created in the imagination, that this imagination is fostered and that people live it out and make it real. It amounts to the same thing for us, whether there were " so-called " or " real " rosicrucians; the substance of their teaching lives and this substance, which is evident in literature, was what I referred to when I said that rosicrucianism is identical with higher alchemy or the hermetic or the royal art. But I think the comparison holds true for the gold and rose-cross societies also, for the spiritual scope of this new edition is the same as that of the old order, except that, as in the fate of all subtile things, it was misunderstood by the majority. There were not lacking attempts to dissuade people from their errors. In the rosicrucian notes to the " Kompass der Weisen " (edition of 1782), e.g., " Moreover the object of our guiltless guild is not the making of gold. . . . Rather we remove the erroneous opinion from them [the disciples] in so far as they are infected with it, even on the first step of the temple of wisdom. They are earnestly enjoined against these errors and that they must seek the kingdom of God and his righteousness." Also through all kinds of reforms we seek to set the wayfarer on the right path that leads to the original

ideal.   It appears that the alchemistic preparation
of the "work" is available only for the smallest
circles.   The multitude is blinded.

"Where do the Scottish masters stay?"

"Quite near the sun."

"Why?"

"Because they can stand it."

# THE PROBLEM OF MULTIPLE INTERPRETATION

AFTER what has been said it is clear that the Parable contains instruction in the sense of the higher alchemy. Whoever has attentively read this 4th chapter will certainly be in a position to understand the parable, in large part, in a hermetic sense. I do not wish to develop this interpretation now, for to a certain extent it develops itself without further effort, and what goes beyond that can be treated only in the second part of this volume. I shall limit myself now to a few suggestions.

In regard to the external setting of the parable as a piece of rosicrucian literature, we must remember that it was published in 1788, the time of the later gold- and rose-cross societies, and in a book whose theosophic and religious character is seen in all the figures contained in it as well as in the greater part of the text. It is continually reiterated that gold is not common gold but our gold, that the stone is a spiritual stone (Jesus Christ), etc. The creation of the world, the religious duty of mankind, the mystic path to the experiencing of divinity — all is represented in detailed pictures with predominantly chemical symbolism. This higher conception of alchemy, that corresponds throughout to the ideal of the so-called old or true rosicrucian, does not prevent the editor from believing in the possibility of miraculous gifts which are to be gained through the hermetic art.    Many parts of the book make us suspect a certain naïveté that may go several degrees beyond the simplicity required for religious development.

As for the origin of the parable there are two possibilities.    Either the editor is himself the author and as such retires into the background, while he acts as collector of old rosicrucian manuscripts, that he now in publishing, discloses to amateurs in the art, or the editor is merely editor.    In either case the obligation remains to interpret the parable hermetically.    The educational purpose of the editor is established.    If he is himself the author, he himself has clothed his teachings in the images of the parable.    If, on the contrary, the author is some one else (either a contemporary and so ☉ R C. ✠, or an old hermetic philosopher, Fr. R. C.), the editor has found in the piece edited by him a subject suitable to his purpose, a material that voices his doctrines.    We can evidently also rest satisfied, in order to evade the question of authorship, that the writing itself gets its own character from the hermetic interpretations, and shows in detail its correspondingly theosophic material.    Nevertheless I desire to show the directing hand of the collector and editor.

Several controlling elements pointing toward a hermetic theosophic interpretation, which the reader probably looks for in the parable, may be shown if I mention the ethical purposes that here and there emerge in our psychoanalytic interpretation of the parable. I might remind the reader that the wanderer is a killer of dragons like St. George; the holy Mary is represented standing over a dragon; also under the Buddha enthroned upon a lotus flower, there curls not infrequently a vanquished dragon; etc. I might mention the religious symbolism of the narrow path that leads to the true life. Many occurrences in the parable are to be conceived as trials, and we can see the wanderer overcome the elemental world (Nature triumphs over Nature), wherein he is proved by all four elements and comes off victorious from all tests. The fight with the lion in the den can be regarded as a world test, the walk on the cloud capped wall (like the flying up in the vessel) as an air test, the mill episode (and the flood in the vessel) as a water ordeal, and the stay in the heated vessel as a fire ordeal. The old miller is God, the ten mill wheels are the ten commandments, and likewise the ten Sephiroth that create the whole world. We are also reminded of the Ophanim (wheels, a class of angels).

Several particulars suggest the admission of the seeker into a hermetic fraternity, which, as far as I am concerned, might be called rosicrucian. There was also among the cabbalists, as apparently is shown by Reuchlin (De Vero Mirifico), an initiation into a mystery. Fludd (in his Tractatus theologo-philo-

sophicus de vita, morte et resurrectione, Chap. XVI)
apostrophizes the rosicrucians: "With open eyes I
saw from your brief answer to two men whom you
intended, at the exhortation of the Holy Ghost, to
choose to your cloister or house, that you possessed
the same knowledge of the true mystery and the same
keys of knowledge that unlock the Paradise of Joy,
as the patriarchs and prophets of holy scripture possess." And in another place, "Believe that your
(the R C. ✠) palace or abode is situated at the con-
fines of the earthly paradise [locus voluptatis terres-
tris]. . . ." In our parable it is a paradise of joy
[pratum felicitatis] where the wanderer meets the
company into which he desires admission. He must
undergo examinations like every neophyte. The
collegium sapientiae of the parable refers to the
rosicrucian Collegium Sancti Spiritus, which is
actually named in another passage of the book that
contains the parable.

The blood of the lion, which the wanderer gets
by cutting him up, refers to the rose-colored blood
of the cross that we gain through deep digging and
hammering. The wanderer picks roses and puts
them in his hat, a mark of honor. The master is
generally seen provided with a hat in the old pictures. "Rose garden" (the garden of the parable
is quadrangular) was a name applied apparently to
alchemistic lodges. The philosophical work itself
is compared to the rose; the white rose is the white
tincture, the red rose is the red tincture (different
degrees of completion that follow the degrees of

black). They are plucked in the " alchemistic paradise," but one must set about it in obedience to nature. Basilius Valentinus in the third of his twelve
keys writes of the great magisterium: " So whoever wishes to compare our incombustible sulphur
of all the wise men, must first take heed for himself,
that he look for our sulphur in one who is inwardly
incombustible; which cannot occur unless the salt sea
has swallowed the corpse and completely cast it up
again. Then raise it in its degree, so that it sur-
pass in brilliance all the stars of heaven, and become
in its nature as rich in blood, as the pelican when he
wounds himself in his breast, so that his young may
be well nourished without malady to his body, and
can eat of his blood. [The pelican possesses under
its bill a great pouch in which he can preserve food,
principally fish. If he regurgitates the food out of
his crop to feed his young he rests his bill against
his breast. That gave rise to the belief that it tore
open its breast in order to feed its young with its
blood. From early times the pelican is therefore
used as a symbol of Christ, who shed his blood for
mankind. The alchemists represented the philosopher's stone, the red tincture, as a pelican; for by
its projection on the baser metals it sacrificed itself
and, as it were, gave its blood to tincture them.
The Christian and the hermetic symbolism are concurrent as in higher sense the stone Christ, i.e., the
Messiah, is on our hearts.] That is the rose of our
master with color of scarlet and red dragon's blood,
written of by many, also the purple mantle of the

highest commanders in our art, with which the Queen of Salvation is clothed, and by which all the poor metals can be warmed. Keep well this mantle of honor."

It is interesting that dream parallels can support us in both directions on the path of hermetic interpretation. I have in the second section of this volume reported the " dream of the Flying Post." I must now complete its interpretation. Stekel writes (l. c., p. 399): "If we examine the birth and uterus phantasies, Mr. X. Z., the dreamer, turns out to be a base criminal. He struggles with conscious murder ideas. He is afraid he may kill his uncle or his mother. He is very pious. But his soul is black as the coal-dust-strewn street. His evil thoughts ( the homosexual) pursue him. He enters the mill. It is God's mill that grinds slowly but surely. His weight (his burden of sin) drives the mill. He is expelled. He enters the Flying Post. It is the post that unites heaven and earth. He is to pay, i.e., do penance for his sins. His sins are erotic (three heller = the genitals). His sins and misdeeds stink before heaven (dirty feet). The conductor is death. . . . The wheel room refers to the wheel of criminals. The water is blood." The perilous situation in the dream, God's mill, the blackness, the water or blood, which are their analogues, are found in the parable without further reference being necessary. Especially would I select the unusual detail of the stinking, dirty feet, for which probably no one would see any association in the

parable.   It is found in the episode of the rotting
of the bridal pair in the receptacle.   It is expressly
stated that the putrefying corpses (i.e., the disinte-
grating sinful bodies of men in the theosophic work)
stink.   The opposite is the odor of sanctity.   Actually this opposition recurs frequently in hermetic
manuals.   The conductor in the dream is described
hermetically as a messenger of heaven ☿, Hermes,
conveyor of souls.   His first appearance in the life
of man is conscience.   This causes our sins, which
would be otherwise indifferent, to stink.   In alchemy
the substances stink on their dissolution in mercuri-
ous purifying liquid.   Only later does the agreeable
fragrance appear.

   If we find on the one hand that the parable appears as a hermetic writing, which allows us to develop theosophical principles from its chemical
analogues, on the other hand the psychoanalytic
interpretation is not thereby shaken.   Consequently
the question arises for us how it is possible to give
several interpretations of a long series of symbols
that stand in complete opposition.   [If we were
concerned with individual symbols merely, the matter would not be at all extraordinary.]   Our research has shown that they are possible.   The psychoanalytic interpretation brings to view elements of
a purposeless and irrational life of impulse, which
works out its fury in the phantasies of the parable;
and now the analysis of hermetic writings shows us
that the parable, like all deep alchemistic books, is
an introduction to a mystic religious life,— according

to the degree of clearness with which the ideas hovered before the author. For just as the psychoanalytically derived meaning of the phantasies does not occur to him, so possibly even the mystical way on which he must travel must have appeared only hazily before him. So no matter what degree of clearness the subjective experience may have had from the author's point of view, we have for the solution of our own problem, to stick to the given object and to the possibilities of interpretation that are so extraordinarily coherent.

The interpretations are really three; the psychoanalytic, which leads us to the depths of the impulsive life; then the vividly contrasting hermetic religious one, which, as it were, leads us up to high ideals and which I shall call shortly the anagogic; and third, the chemical (natural philosophical), which, so to speak, lies midway and, in contrast to the two others, appears ethically indifferent. The third meaning of this work of imagination lies in different relations half way between the psychoanalytic and the anagogic, and can, as alchemistic literature shows, be conceived as the bearer of the anagogic.

The parable may serve as an academic illustration for the entire hermetic (philosophy). The problem of multiple interpretation is quite universal, in the sense namely that one encounters it everywhere where the imagination is creatively active. So our study opens wide fields and art and mythology especially appear to invite us. I will depart as little as possible, however, from the province chosen as an example, i.e., alchemy. But in two fables I shall work out the problem of multiple interpretation. In the choice of the fables I am influenced by the fact that a psychoanalytic elaboration (Rank's) lies ready to hand, and that both are subjected to an anagogic interpretation by Hitchcock, who wrote the book on alchemy. This enables me to take the matter up briefly because I can simply refer to the detailed treatment in the above mentioned books. The two stories belong to Grimm's collection and are called the Six Swans, and the Three Feathers. (K. H. M., Nos. 49 and 63.)

Rank (Lohenginsage) connects the story of the six swans and numerous similar stories with the knight of the swan saga. It is shown that the mythical contents of all these narratives have at bottom those elemental forces of the impulse life that we have found in the parable, and that they are specially founded on family conflicts, i.e., on those uncontrolled love and hate motives that come out in their crassest form in the neurotic as his (phantasied) " family romance." To this family romance belongs, among others, incest in different forms, the illicit love for the mother, the rescuing of the mother from peril, the rescuing of the father, the wish to be the father, etc., phantasies whose meaning is explained in the writings of Freud and Rank (Myth of the Birth of the Hero [1]). According to Hitch-

---

[1] Nervous and Mental Disease Monograph Series. Tr. by Jelliffe.

cock, on the contrary, the same story tells of a man
who in the decline of life falls into error, takes the
sin to his heart, but then, counseled by his conscience,
seeks his better self and completes the (alchemic-
creative) work of the six days. (Hitchcock, Red
Book.)

It is incontestable that there is, besides the psychoanalytic and anagogic interpretation of this tale (and
almost all others), a nature mythological and in the
special sense, an astronomical interpretation. Significant indications of this are the seven children and
the seven years, the sewing of clothes made of star
flowers, the lack of an arm as in the case of Marduk,
and the corresponding heroes of astral myths, and
many others. One of the seven is particularly distinguished like the sun among the so-called planets.
The ethically indifferent meaning of the tale alongside of the psychoanalytic and the anagogic corresponds to the chemical contents of the hermetic writings. As object of the indifferent meaning there
always stands the natural science content of the
spirit's creation. There is generally a certain relationship between the astronomical and the alchemistic meanings. It is now well known that alchemy
was influenced by astrology, that the seven metals
correspond to the seven planets, that, as the sun is
distinguished among the planets, so is gold among
the metals; and as in astrology combustion takes
place in heaven, so it occurs also in the alembic of
the alchemists. And the fact that the sun maiden
at the end of the story releases her six planet broth-

ers, sounds exactly as when the tincturing power of
gold at the end of six days perfects the six imperfect
metals and makes the ill, well.

In the second story I will emphasize to a somewhat greater degree the opposition of the two contrasting interpretations (psychoanalytic and anagogic), as I must return to it again.   The story is
suited to a detailed treatment on account of its brevity.   I will first present it.

There was once a king who had three sons, two
of whom were clever and shrewd, but the third did
not talk much, was simple and was merely called
the Simpleton.   When the king grew old and feeble
and expected his end, he did not know which one of
his sons should inherit the kingdom after him.   So
he said to them, " Go forth, and whoever brings me
the finest carpet shall be king after my death."   And
lest there be any disagreement among them, he led
them before his castle, blew three feathers into the
air, and said:   " As they fly, so shall you go."   One
flew towards the east, the other towards the west, the
third, however, flew straight ahead, but flying only
a short distance soon fell to earth.   Now one
brother went to the right, the other went to the left,
and they laughed at Simpleton, who had to stay
with the third feather where it had fallen.

Simpleton sat down and was sad.   Suddenly he
noticed that near the feather lay a trap door.   He
raised it, found a stairway, and went down.   Then
he came before another door, knocked and listened,
while inside a voice called:

> "Maiden green and small,
> Shrunken old crone,
> Old crone's little dog,
> Crone here and there,
> Let us see quickly who is out there."

The door opened and he saw a big fat toad and round about her a crowd of little toads. The fat toad asked what his wish was. He answered, "I should have liked the most beautiful and finest carpet." Then she called a young one and said:

> "Maiden green and small,
> Shrunken old crone,
> Crone's little dog,
> Crone here and there,
> Fetch here the big box."

The young toad brought the box and the fat toad opened it and gave Simpleton a carpet from it, so beautiful and so fine as up above on the earth could not have been woven. Then he thanked her and climbed up again.

The two others had, however, considered their youngest brother so weak-minded that they believed that he would not find and bring anything back. "Why should we take so much trouble," said they, and took from the back of the first shepherd's wife that met them her coarse shawl and carried it home to the king. At the same time Simpleton returned and brought his beautiful carpet, and when the king saw it he was astonished and said: "If justice must be done, the kingdom belongs to the youngest."

But the two others gave their father no peace, and said that it was impossible that Simpleton, who lacked understanding in all things, could be a king, and begged him to make a new condition. Then the father said, " The one that brings me the most beautiful ring shall be king," led the three brothers out and blew three feathers into the air for them to follow. The two oldest again went east and west, and Simpleton's feather flew straight ahead and fell down near the door in the earth. So he went down again to the fat toad and told her that he needed the most beautiful ring. She immediately had her big box fetched and from it gave him a ring that glittered with jewels and was more beautiful than any goldsmith upon the earth could have made. The two eldest laughed about Simpleton, who was going to look for a gold ring, but they took no trouble, and knocked the pin out of an old wagon ring and brought the ring to the king. But when Simpleton showed his gold ring the father again said, " The kingdom belongs to him." The two eldest did not cease importuning the king till he made a third condition and declared that the kingdom should go to the one that brought home the fairest woman. Again he blew the three feathers into the air and they flew as before.

So Simpleton without more ado went down to the fat toad and said, " I have to take home the fairest woman." " The fairest woman, hey? She is not right here, but none the less you shall have her." She gave him a hollowed out carrot to which were harnessed six little mice. Then Simpleton sadly said, " What shall I do with it ? "  The toad replied, " Just put one of my little toads in it." So he took one by chance from the circle and put it in the yellow carriage, but hardly had she taken her seat when she became a surpassingly beautiful maiden, the carrot a coach, and the six little mice, horses. So he kissed the maiden, drove away with the horses and took them to the king. His brothers came afterwards. They had not taken any trouble to find a fair lady but had brought the first good looking peasant woman. As the king looked at them he said, " The youngest gets the kingdom after my death." But the two oldest deafened the king's ears with their outcry: " We cannot allow the Simpleton to be king," and gained his consent that the one whose woman should jump through a ring that hung in the middle of the room should have the preference. They thought, " The peasant women can do it easily, they are strong enough, but the delicate miss will jump herself to death." The old king consented to this also. So the two peasant women jumped, even jumped through the ring, but were so clumsy that they fell and broke their awkward arms and legs. Then the beautiful woman whom Simpleton had brought leaped through as easily as a roe, and all opposition had to cease. So he received the crown and ruled long and wisely.

I offer first a neat psychoanalytic interpretation of this narrative. Like the dream, the fairy tale is regularly a phantastic fulfillment of wishes, and, of such indeed, as we realize, but which life does not satisfy, as well as of such as we are hardly aware of in consciousness, and would not entertain if we knew them clearly. Reality denies much, especially to the weak, or to those who feel themselves weak, or who have a smaller capacity for work in the struggle for existence in relation to their fellow men. The efficient person accomplishes in his life what he wishes, the wishes of the weak remain unfulfilled, and for this reason the weak, or whoever in comparison with the magnitude of his desires, thinks himself weak, avails himself of the phantastic wish fulfillment. He desires to attain the unattainable at least in imagination. This is the psychological reason why so many fairy stories are composed from the standpoint of the weak, so that the experiencing Ego of the fairy tale, the hero, is a simpleton, the smallest or the weakest or the youngest one who is oppressed, etc. The hero of the foregoing tale is a simpleton and the youngest. In his phantasy, that is, in the story, he stamps his brothers, who are in real life more efficient, and whom he envies, as malicious, disagreeable characters. (In real life we can generally observe how suspicious are, for instance, physically deformed people. Their sensitiveness is well known.) Like the fox to whom the grapes are sour, he declares that what his stronger fellows accomplish is bad, their performance of their duty defective, and their aims contemptible, especially in the sexual sphere, where he feels himself openly most injured. The tale treats specifically from the outset the conquest of a woman.   The carpet, the ring, are female symbols, the first is the body of the woman, the ring is the vagina (Greek kteis = comb = pudenda muliebria).   (The carpet is still more specifically marked as a female symbol in that the brothers take it from the body of a shepherdess.   Shepherdess — a coarse " rag "— coarse " cloth "— in contrast to the fine carpet of the hero.)

The simpleton is one who does not like much work.   When he also ascribes negligence to his brothers he betrays to us his own nature, in that his " feather," i.e., himself, does not go far, while his brothers' feathers go some distance.   In order to invalidate this view of himself the distribution of the feathers is put off on chance, as if to a higher determining power.   This has always been a favorite excuse with lazy and inefficient people.

One of the means of consoling himself for the unattainableness of his wishes is the belief in miracles. (Cf. my work on Phantasy and Mythos.)   The simpleton gains his advantage in a miraculous manner; roasted pigeons fly into his mouth.

In his erotic enterprises he sticks to his own immediate neighborhood.   He clearly bears within himself an Imago that holds him fast.   [This is an image, withdrawn from consciousness and consequently indestructible, of the object of one's earliest passion, which continues to operate as a strongly affective complex, and takes hold upon life with a formative effect.   The most powerful Imagos are those of the parents.   Here naturally the mother imago comes to view, which later takes a position in the center of the love life (namely the choice of object).] Whither does he turn for his journey of conquest? Into the earth.   The earth is the mother as a familiar symbol language teaches us.   Trap door, box, subterranean holes, suggest a womb phantasy.   The toad frequently appears with the significance of the uterus, harmonizing with the situation that the tale presents.   (On the contrary frog is usually penis.) The toad's big box (= mother) is also the womb. From it indeed the female symbols, in this connection, sisters, are produced for the simpleton.   The box is, however, also the domestic cupboard,— food closet, parcel, bandbox, chamber, bowl, etc.,— from which the good mother hands out tasty gifts, toys, etc.   Just as the father in childish phantasy can do anything, so the mother has a box out of which she takes all kinds of good gifts for the children.   Down among the toads an ideal family episode is enacted. The mother's inexhaustible box (with the double meaning) even delivers the desired woman for the simpleton.

The woman — for whom?   Doubtless for the simpleton, psychologically.   The tale says for the king, because the female symbols, carpet, ring, the king desires for himself, in so many words, and the inference is that the woman also belongs to him. The conclusion of the tale, however, turns out true to the psychological situation, as it does away with the king and lets the simpleton live on, apparently with the same woman.   It is clear as day that the simpleton identifies himself with his father, places himself in his place. The image, which possesses him from the first is the father's woman, the mother. And the father's death — that is considerately ignored — which brings queen and crown, is a wish of the simpleton. So again we find ourselves at the center of the Œdipus complex. As mother-substitute figures the sister, one of the little toads.

We have regarded the story first from the point of view of the inefficiency of the hero, and have thereupon stumbled upon erotic relations, finally upon the Œdipus complex. The psychological connection results from the fact that those images on which the Œdipus complex is constructed appear calculated to produce an inefficiency in the erotic life.

The anagogic interpretation of Hitchcock (l. c., pp. 175 ff.) is as follows, though somewhat abridged:

The king plainly means man. He has three sons; he is an image of the Trinity, which in the sense of our presentation we shall think of as body, soul and spirit. Two of the sons were wise in the worldly sense, but the third, who represents spirit and in the primitive form, is called conscience, is simple in order to typify the straight and narrow path of truth. The spirit leads in sacred silence those who meekly follow it and dies in a mystical sense if it is denied, or else appears in other forms in order to pursue the soul with the ghosts of murdered virtues. Man is, as it were, in doubt concerning the principle to which the highest leadership in life is due. " Go

forth and whoever brings me the finest carpet shall
be king after my death." The carpet is something
on which one walks or stands, here representing the
best way of life according to Isaiah xxx, 21. " This
is the way, walk ye in it, when ye turn to the right
hand and when ye turn to the left."

The three feathers are, of course, the three principles. Two of them move at once in opposite directions [towards the east and towards the west, as
many writers on alchemy represent the two principles or breaths, anima and corpus or ☉ and ☽ ]
and so come even at the outset away from the right
path. The third, symbol of the spirit, flies straight
forward and has not far to its end, for simple is
the way to the inner life. And so the spirit will
speak to us if we follow its voice, at first quite a
faint voice: " But the word is very nigh unto thee,
in thy mouth and in thy heart that thou mayst do it."
(Deuteronomy xxx, 14.) Yet the soul is not free
from sadness, as the man stands still on the lower
steps of the ladder that leads up into eternal life.
Simpleton is troubled in his heart and in the humility
of this affliction he discovers " all at once " a secret
door, which shows him the entrance into the mystical
life. The door is on the surface of the earth, in
abasement, as the third feather determined it in
advance. As Simpleton discreetly obeyed it, he
strolled along the path that the door opened for him.
Three steps, three fundamental forces. So Christ
had to descend before he could rise. The hero of
the story knocks as Christ knocks in the gospel (i.e.,

on the inner door, contrasted with the law of Moses, the outer door). The big toad with her little ones in a circle about her signifies the great mother nature and her creatures, which surround her in a circle; in a circle, for nature always returns upon herself in a cycle. Simpleton gets the most beautiful carpet.

The other two beings that we call understanding and feelings (sun and moon of the hermetic writings) look without, instead of seeking the way within; so it comes to pass that they take the first best coarse cloths.

To bring the most beautiful ring is to bring truth, which like a ring has neither beginning nor end. Understanding and feeling go in different directions, the simpleton waits meekly by the door that leads to the interior of the great mother. [The appearance of this conception in the anagogic interpretation is also important.]

In the third test, the search for " the fairest woman," the crown of life, conceived exoterically as well as esoterically, the carrot represents the vegetative life (body, the natural man), and the six mice that draw it are our old friends the six swans or virtues, and the highest of these compassion — or love — goes as the enthroned queen in the carriage. The uninitiated man is almost in doubt and asks, " What shall I do with a carrot? " Yet the great mother replies, as it were, " Take one of my fundamental forces." And what do we see then? The toad becomes a beautiful maiden, etc. The man now all at once realizes how fearfully and wonderfully he is made. Filled with reverence of himself he is ready to cry, " Not my will but thine be done."

Still another test remains. We must all go through a sort of mystical ring, which hangs in the hall (of learning). Only one in the whole universe is in a condition to accomplish it, to endure it without injury. The beautiful delicate maid with the miraculous gift is the spirit [spiritus or ☿ of alchemy].

We shall add that the two interpretations externally contradict each other, although each exhibits a faultless finality. I should note that I have limited myself to the briefest exposition; in a further working out of the analysis the two expositions can be much more closely identified with the motives of the story.

First, then, the question arises, how one and the same series of images can harmonize several mutually exclusive interpretations (problem of multiple interpretation) ; yet we have discovered in the parable three practically equivalent schemes of interpretation, the psychoanalytic, the chemical (scientific), and the anagogic. Secondly, the question presents itself more particularly how can two so antithetic meanings as the psychoanalytic and the anagogic exist side by side.

# III
# SYNTHETIC PART

## Section I

## INTROVERSION AND REGENERATION

### A. INTROVERSION AND INTRO-DETERMINATION

THE multiple interpretation of works of fantasy has become our problem, and the diametrical opposition of the psychoanalytic and the anagogic interpretation has particularly struck us. The question now apparently becomes more complicated if I show that the psychoanalytic interpretation contains an analogue that we must take into consideration. The analogue is presented by the remarkable coexistence of symbolism of material and functional categories in the same work of imagination. In order to make myself intelligible, I must first of all explain what these categories are.

In division 2 of the introductory part we have seen that the imagination shows a predilection for symbolic forms of expression, proportionately greater indeed, the more dreamlike it is. Now by this symbolism as we observe most clearly in hypnagogic (half dreaming) hallucinations and in dreams, three different groups of objects are represented.

I. Thought contents, imagination contents, in brief, the contents or objects of thinking and imagining, the material of thought whether it be conscious or unconscious.

233

II. The condition, activity, structure of the psyche, the way and manner that it functions and feels, the method of functioning of the psyche, whether it be conscious or unconscious.

III. Somatic processes (bodily stimulations). This third sort of objects is closely coördinated with the other two.    It is not capable of interesting us in the present connection so we pass it by.

Therefore we arrive at two categories in which we can enroll all symbolizing works of the imagination, the material and the functional.

I. The material category is characterized by its representation of thought contents, i.e., of contents that are worked out in a train of thoughts (arranged thought, imagined), whether they are mere images or groups of images, concepts that are somewhat drawn out into comparisons and definition processes, or indeed judgments, trains of reasoning, which serve as analytic or synthetic operations, etc.    Since, as we know, the phantasies (dreams, reveries, even poems) are mostly inspired by wishes, it will prove frequently the case that the contents symbolically contained in them are wish images, i.e., the imagined experiencing of gratification.

II. The functional category is characterized by the fact that the condition, structure or capacity for work of the individual consciousness (or the psychic apparatus) is itself portrayed.    It is termed functional because it has nothing to do with the material or the contents of the act of thinking, but applies merely to manner and method in which consciousness functions (rapid, slow, easy, hard, obstructed, careless, joyful, forced; fruitless, successful; disunited, split into complexes, united, interchangeable, troubled, etc.). [It is immaterial whether these are conscious or unconscious. Thinking must be taken here in the widest possible sense. It means here all psychic processes that can have anything as an " object."]

Two typical examples will enable us at once clearly to understand the two categories and keep them separate.

A. Material Symbolism.— Conditions. In a drowsy state I reflect upon the nature of the judgments that are transsubjectively (= for all men) valid. All at once the thread of the abstract thought is broken and autosymbolically in the place of it is presented the following hypnagogic hallucination:

Symbol. An enormous circle, or transparent sphere, floats in the air and men are putting their heads into this circle.

Interpretation. In this symbol everything that I was thinking of is expressed. The validity of the transsubjective concerns all men without exception; the circle goes through all the heads. This validity must have its cause in something common to all. The heads all belong to the same apparently homogeneous sphere. Not all judgments are transsubjective; with their bodies and limbs men are outside of and under the sphere and stand on the earth as separate individuals.

B. Functional Symbolism.— Conditions.   Dreamy state as above.   I reflect upon something or other, and yet in allowing myself to stray into bypaths of thought, I am diverted from my peculiar theme. When I want to get back the autosymbolic phenomenon appears.

Symbol.   I am climbing mountains.   The nearer mountains shut out my view of the more distant ones, from which I have come and to which I should like to return.

Meaning.   I have got off the track.   I have ventured too high and the ideas that I have entertained shut out my starting point like the mountains.

To the material category belongs, for example, the meaning of the strawberry dream explained in the second part of the introductory chapter.   Strawberry picking is a symbol for an imaged wish gratification (sexual intercourse), and so for an image content.   The symbolism is therefore a material one.   The greatly preponderating part of psychoanalytic dream literature is occupied with interpretation according to material categories.

To the functional categories belong, for example, the symbolism of falling asleep and waking up, which I have mentioned in the second part in connection with the interpretation of the parable.

The two categories of symbolism, if they never did anything but parallel each other, would afford us no analogues for our problem of double meaning. Now the cases, however, are extremely rare where there is only functional or only material symbolism; the rule is an intimate interweaving of both. To be sure, one is frequently more emphasized than the other or more easily accessible, but we can generally find cases where long contexts of images are susceptible of material as well as functional interpretation, alike in detail and continuity of connection.

The following may serve as a very simple case in point. Lying one evening in bed and exhausted and about to fall asleep, I devoted my thoughts to the laborious progress of the human spirit in the dim transcendant province of the mothers-problem. (Faust, Part II.) More and more sleepy and ever less able to retain my thoughts, I saw suddenly with the vividness of an illusion a dream image. I stood on a lonely stone pier extending far into a dark sea. The waters of the sea blended at the horizon with an equally dark-toned mysterious, heavy air. The overpowering force of this tangible picture aroused me from my half sleeping state, and I at once recognized that the image, so nearly an hallucination, was but a visibly symbolic embodiment of my thought content that had been allowed to lapse as a result of my fatigue. The symbol is easily recognized as such. The extension into the dark sea corresponds to the pushing on into a dark problem. The blending of atmosphere and water, the imperceptible gradation from one to the other means that with the " mothers " (as Mephistopheles pictures it) all times and places are fused, that there we have no boundaries between a " here " and a " there," an " above " and a " below," and for this reason Mephistopheles can say to Faust on his departure,

" Plunge then.— I could as well say soar."

We see therefore between the visualized image and the thought content, which is, as it were, represented by it, a number of relations. The whole image resolves itself insofar as it has characteristic features, almost entirely into such elements as are most closely related to the thought content. Apart from these connections of the material category, the image represents also my momentary psychic condition (transition to sleep). Whoever is going to sleep is, as it were, in the mental state of sinking into a dark sea. (The sinking into water or darkness, entrance into a forest, etc., are frequently-occurring threshold symbols.) The clearness of ideas vanishes there and everything melts together just as did the water and the atmosphere in the image.

This example is but to illustrate; it is in itself much too slight and simple to make any striking revelation of the remarkable interlacing of the two kinds of symbolism. I refer to my studies on symbolism and on dreams in the bibliography. Exhaustive treatment at this point would lead us too far afield. Let us rest satisfied then with the facts that the psychoanalysts simultaneously deal with two fundamentally different lines of interpretation in a product of the phantasy (dream, etc.), quite apart from the multiple determinants which they can find within the material as well as in the functional categories; both lines of interpretation are supplied by the same fabric of images, indeed often by the same elements of this image fabric. This context therefore must have been sought out artfully enough by the creative unconscious to answer the double requirement.

The coexistence of the material meaning with the functional is not entirely puzzling to the student of psychoanalysis. Two facts must be kept in mind throughout.

In the first place, we are acquainted with the principle of multiple determination or condensation. The multiplicity of the dimly moving latent dream thoughts condenses into a few clear dream forms or symbols, so that one symbol continually, as it were, appears as the representative of several ideas, and is therefore interpretable in several ways. That it should be susceptible of more than one interpretation can cause no surprise because the fundamental significance (the latent thoughts) were the very ones that, by association, caused the selection of the symbols from an infinite series of possibilities. In the shaping of the dream, and therefore in the unconscious dream work, only such pictorial elements could penetrate into consciousness as satisfied the requirements of the multiple determination. The principle of multiple determination is valid not only within the material and the functional categories, but makes the fusion of both in the symbol in question to some extent intelligible. Elements of both categories take an active part in the choice of the symbol.   On the one hand, a number of affects press on towards the symbolic representation of objects to which they direct themselves (objects of love, hate, etc.).   On the other hand, the psyche takes cognizance of its own impulses, play of affects, etc., and this perception will gain representation.   Both impulses take part in the choice of those symbols which thrust themselves into the nascent consciousness of phantasy, and so the dream, like the poem, etc., besides the symbolism of the wish tendencies (material categories) that animate them, bears the stamp of the psychic authorship (functional category) of the dreamer or the author.   [Ferenczi defends the view for the myth also that the material symbolism must coincide with the functional (Imago I, p. 283).]

Secondly, it has been shown in recent times in psychoanalytic studies that symbols which were originally material pass over to functional use.   If we thoroughly analyze for a sufficient time the dreams of a person we shall find that certain symbols which at first probably appeared only incidentally to signify some idea content, wish content, etc., return and become a persistent or typical form.   And the more such a typical form is established and is impressed, the farther it is removed from its first ephemeral meaning, and the more it becomes a symbolic representative of a whole group of similar experiences, a spiritual capital, so to speak, till finally we can regard it simply as the representative of a spiritual current (love, hate, tendency to frivolity, to cruelty, to anxiety, etc.). What has been accomplished there is a transition from the material to the functional on the path of a determination inward or intro-determination (verinnerlichung) as I shall call it. Later I shall have more to say about intro-determination. For the present this may suffice for the understanding, that the material and the functional symbolism, in spite of their at first apparently fundamental difference, are essentially related in some way, which is illuminated by the process of intro-determination.

The analogue of the problem of multiple interpretation unfolded in the preceding section is shown to be a question that can be easily answered. And we would bring our problem to a generally satisfactory position if we succeeded in showing that the anagogic interpretation, whose alignment with the psychoanalytic seemed so impracticable, is a form of functional interpretation, or at least related to it. In this case it would be at once comprehensible how a product of the imagination harmonizes with several expositions (problem of multiple interpretation); because this variety of sense had already operated in the selection of the symbol and indeed, in those cases as well where we did not at first sight suspect the coöperation of the anagogic thoughts; secondly, the anagogic and the psychoanalytic interpretations are somehow reconciled to each other, whereby possibly also the position of the natural science interpretation can be made somewhat clearer.

The possibility that the anagogic has some part

in the creation of the functional, will be brought
nearer by the fact that our previously offered anagogic expositions (fairy tales, parabola) markedly
resemble functional interpretations.    In the tale of
the six swans Hitchcock explains the reception of
the maiden into the castle as the reception of sin into
the heart; the seven children are the seven virtues
(consequently    spiritual    tendencies).    The    small
maiden is conscience, the tissues are processes of
thought.    In the story of the three feathers, again,
one son is conscience; the secret door is the entrance
to the inner life, to spiritual absorption, the three
feathers are spiritual tendencies, etc.    In the dream
of the " flying post " conscience appears as the conductor.    The " Mills of God," which psychologically also represents conscience, the more strikingly
because the burden of sin, guilty feeling, drives them,
also appear in the parable.    The lion or the dragon
which must be overcome on the mystic path is again
a spiritual force.    The approximation to the functional category is not to be denied.    Processes that
show an interplay of spiritual powers are symbolically represented there.    But we are at once struck
with a difference.    The true functional phenomenon,
as I have so far described it, pictures the actual
psychic state or process; the anagogic image appears
on the contrary to point to a state or process that
is to be experienced in the future.    We shall pass
over for a time the last topic, which will not, however, be forgotten, and turn to the question as to
the point on which the anagogic and the functional

interpretations can best be brought together. This point appears to me to be introversion, first because it is related to the previously mentioned intro-determination, and second, because it is familiar to psychoanalysis and is of great importance in anagogic method.

The term " introversion " comes from C. G. Jung. It means sinking into one's own soul; the withdrawal of interest from the outer world; the seeking for joys that can be afforded by the inner world. The psychology of the neuroses has led to the concept of introversion, a province, therefore, which principally treats of morbid forms and functions of introversion. The sinking of oneself into one's own soul also appears exactly as a morbid losing of oneself in it. We can speak of introversion neuroses. Jung regards dementia precox as an introversion neurosis. Freud, who has adopted the concept of introversion [with some restrictions] regards the introversion of the libido as a regular and necessary precondition of every psychoneurosis. Jung (Jb. ps. F., III, p. 159) speaks of " certain mental disturbances [he means dementia precox] which are induced by the fact that the patients retire more and more from reality, sink into their phantasy, whereby in proportion as reality loses its force, the inner world takes on a reality and determining power." We may also define introversion as a resignation of the joys of the outer world (probably unattainable or become troubled) and a seeking for the libido sources in one's own ego. So we see how generally self-chastisement, introversion and autoerotism are connected.

The turning away from the outer world and turning in to the inner, is required by all those methods which lead to intensive exercise of religion and a mystic life. The experts in mysteries provide for opportunities that should encourage introversion. Cloisters and churches are institutions of introversion. The symbolism of religious doctrine and rite is full of images of introversion, which is, in short, one of the most important presuppositions of mysticism.

Religious and mythical symbolism has countless images for introversion; e.g., dying, going down, subterranean crypts, vaults, dark temples, into the underworld, hell, the sea, etc.; being swallowed by a monster or a fish (as Jonah), stay in the wilderness, etc. The symbols for introversion correspond in large part with those that I have described for going to sleep and waking (threshold symbolism), a fact that can be readily appreciated from their actual similarity. The descent of Faust to the mothers is an introversion symbol. Introversion fulfills here clearly the aim of bringing to reality, i.e., to psychological reality, something that is attainable only by phantasy (world of the past, Helen).

In Jacob Boehme (De Vita Mentali) the disciple says to the master, " How may I attain suprasensuous life, so that I may see God and hear him speak? " The master says, " When you can lift yourself for one moment into that realm where no creature dwell-

eth, you will hear what God speaks." The disciple
says, " Is that near or far? " The master says,
" It is in yourself."

The hermetics often urge retirement, prayer and
meditation, as prerequisites for the work; it is
treated of still more in the hieroglyphic pictures
themselves. The picture of death is already familiar to us from the hermetic writings, but in the technical language there are still other expressions for
introversion, e.g., the shutting up in the receptacle,
the solution in the mercury of the sages, the return
of the substance to its radical condition (by means
of the " radical " or root dampness).

Similar features in our parable are the wandering
in the dense forest, the stay in the lion's den, the
going through the dark passage into the garden, the
being shut up in the prison or, in the language of
alchemy, the receptacle.

Introversion is continually connected with regression. Regression, as may be recalled from the 2d
section of Part I, is a harking back to more primitive psychic activities, from thinking to gazing, from
doing to hallucinating; a striving back towards childhood and the pleasures of childhood. Introversion
accordingly is accompanied by a desire for symbolic
form of expression (the mystical education is carried on in symbols), and causes the infantile imagos
to revive — chiefly the mother image. It was pre-
eminently father and mother who appeared as objects of childish love, as well as of defiance. They
are unique and imperishable, and in the life of adults

there is no difficulty in reawakening and making
active those memories and those imagos. We easily
comprehend the fact that the symbolic aim of the
previously mentioned katabasis always has a maternal character; earth, hole, sea, belly of fish, etc.,
that all are symbols for mother and womb. Regression revives the Œdipus complex with its
thoughts of incest, etc. Regression leads back to all
these relics now done away with in life and repressed.
It actually leads into a sort of underworld, into the
world of titanic wishes, as I have called them. How
far this was the case in the alchemistic parable, I
have fully shown in the psychoanalytic treatment of
it. Here I need merely to refer to the maternal nature of the symbols cited: receptacle, mercury of the
sages ("mother of metals") and radical moisture,
also called " milk " and the like.

Fairy tales have frequently a very pretty functional symbolism for the way in which introversion
leads to the mother imago. Thus the simpleton in
the fairy tale of the feathers comes through the
gate of introversion exactly into the family circle, to
the mother that cares for him. There his love finds
its satisfaction. There he even gets a daughter,
replica of the mother imago, for a wife.

In the parable the wandering in the forest (introversion) is followed by the battle (suggestive of
incest) with the lion (father or mother in their awe
inspiring form) ; the inclusion in the receptacle (introversion) by the accomplishment of the incest.

If it is now clear also that in introversion, as a

result of the regression that is connected with it, visions of "titanic" emotions (incest, separating of parents, etc.) are encountered, yet it has not become in the slightest degree comprehensible how these visions are related to the treatment of anagogic ideas. And that is indeed the question.

We can really understand these striking facts better if we recall what I have said above about the type formation and the intro-determination of the symbols, namely, that symbols can depart from their original narrower meaning and become types for an entire class of experiences whereby an advance is made from the material to the functional meaning. Some examples will elucidate this.

I have observed particularly fine cases of intro-determination in a series of experiments in basin divination (lecanomancy) which I have carried on for several years. Lecanomancy resembles crystal gazing, except that the gazer looks into a basin of water. In the visions of my subject, Lea, typical forms were pictured, which always recurred. Regarded as symbols they were, as subsequent analysis showed, almost all subjected to inward accentuation or intro-determination. Thus, for instance, a black cat appeared. At first it appeared as representative of Lea's grandmother, who was cat-like, malicious and fawning. Later the cat stood for the corresponding traits that she perceived in herself. Above all the cat is the symbol of her grandmother, so the grandmother (or cat) is a mental current of Lea. Frequently there appears in the image a Dyas, sometimes in the shape of a two-headed snake, of two hands, of two feet, or of a woman with two faces, etc. Above all, every antithesis appears to have some external meaning, two men who love each other, etc. So it becomes clear that the common element which finds its most pregnant expression in the double faced woman is the Dyas in itself and that it means bisexuality, psychic hermaphroditism. More than that it is definitely certain that the deepest sense of the symbol means a complete dissociation of Lea's character into two different personalities, one of which may be called the savage and the other the mild. (Lea herself uses the expressions cynical and ideal personality.) In one of the later experiments Lea saw her cynic double vividly personified and spoke in this character, which is closely related to the " black cat." The Dyas in the symbols has the value first of a representation of externals (two lovers, etc.), then as symbols of bisexuality. The sexual Dyas can again be conceived as a symbol or characteristic of a still more general and comprehensive dissociation of the ego. A further symbol and one still more tending towards intro-determination was death. Starting from connections with definite external experiences and ideas of actual death, the meaning of the symbol became more and more spiritual, till it reached the meaning of the fading away of psychic impulses. What died symbolically or had to die was represented by an old man who sacrificed himself after suffering all kinds of fortune. The dying of this old man signified, as

analysis showed, the same thing that we call the
" putting off the old Adam " (turning over a new
leaf). The figure of the old man, originally Lea's
grandfather, then her father, came to have this meaning only after a long process of intro-determination.

A few more examples for typical figures.

In many dreams of a woman analyzed by me
(Pauline, in my treatise Zur Symbolbildung), a cow
appears as a typical image. The alternation of this
cow with more or less definite mother symbols leads
to identification of the cow with the mother. Two
circumstantial dreams that were fully analyzed
showed, however, that the cow and other forms with
which she alternated cannot be translated so correctly by the concept of mother as by that of the
maternal authority and finally still more correctly by
self-criticism or conscience, of which maternal authority is but a type. Children figure in Pauline's
case as a result of various experiences, as typical of
obstacles.

In the case of another dreamer the father stands
in similar relation as the determinant that paralyzes
his resolutions.

The climbing of an ascent, usually a symbol of
coitus (hurrying upward which makes us out of
breath), turns out often in a deeper relation as the
effort to get from the disagreeable things of life to
a place of retreat (lonely attics, etc.), inaccessible
to other persons (= thoughts) ; and now we see
that this deeper meaning appears without prejudice
to the first, for even coitus, like all transport, is only

a special case of flight from the outer life, one of the forms of spiritual oblivion. Hence in part the mythologically and psychopathologically important comparison of intoxication, intoxicating drink and sperm, soma and semen. Ascent = coitus is in this case a type for a quite comprehensive class of experience.

Marcinowski found in his analyses that the father in dream life often was a " symbol of an outlived, obsolete attitude." (Z. Bl. f. Ps., II, 9.)

Other examples of types are the phallus, the sun and other religiously revered objects, if we regard them as does Jung (Wandl. u. Sym., Jb. ps. F., III–IV) as a symbol of the libido. [The concept of which is extended by Jung almost to Schopenhauer's Will.] The typical character of divine personalities is moreover quite clearly emphasized by Jung himself.

The snake, about whose significance as a " negative phallus," etc. [developed in detail by Jung], we shall have more to say, can also be regarded as a typical image. Bull, cow and other animal forms are in mythology as in dreams typical transmutations with unlimited possibility of intro-determination. Dogs are often in dreams the representations of animal propensities. The beast is often " la bête humaine " in the dreamer's own inner life. We have become acquainted with the terrible lions, the bears, etc., as father types; here we get a new perspective which makes clear the one-sidedness of our first conception.

Since psychoanalysis has found acceptation, many of its followers believe they are able to solve, with their work of analysis alone, all the psychological, esthetic and mythological problems that come up. We understand only half of the psychic impulses, as indeed we do all spiritual development, if we look merely at the root. We have to regard not merely whence we come but also whither we go. Then only can the course of the psyche be comprehended, ontogenetically as well as phylogenetically, according to a dynamic scheme as it were.

If we apply this fundamental principle to symbolism there develops therefrom the obligation to keep both visible poles in view, between which the advance of significance, the process of intro-determination is completed. (An externalization is also possible, yet the internalization or intro-determination must be regarded as the normal process.) [It corresponds namely to the process of education and progress of culture. This will soon be cleared up.] To the most general type belong then, without doubt, those symbols or frequently disguised images, concerning which we wondered before, that besides representing " titanic " tendencies, they are fitted to represent the anagogic. The solution of the riddle is found the instant we regard these images as types with a certain degree of intro-determination, as types for a few fundamental forces of the soul, with which we are all endowed, and whose typical symbols are for that reason of general applicability. [I will therefore call these types the human elementary types.] For example, if by psychoanalysis we deduce father and mother, etc., from some of the symbols appearing in dreams, we have in these representations of the psychic images, as the psychoanalyst calls them, in reality derived mere types whose meaning will change according to the ways of viewing them, somewhat as the color of many minerals changes according to the angle at which we hold them to the light. The actual father or mother, the experiences that surrounded them, were the material used in the formation of the types; they were external things even if important, while later the father, etc., emerging as symbol, may have significance as a type of the spiritual power of the very person in question; a spiritual power to be sure, which the person in question feels to be like a father for otherwise the father figure would not be suited for the symbol. And we can go so far as to call this spiritual power a father image. That should not however, mislead us into taking that real person, who in the individual case generally (though not always) has furnished the type, for the real or the most essential. The innermost lies in ourselves and is only fashioned and exercised upon persons of the external world.

So then we get for the typical symbol a double perspective. The types are given, we can look through them forward and backward. In both cases there will be distortions of the image; we shall frequently see projected upon each other, things that do not belong together, we shall perceive convergences at vanishing points which are to be ascribed only to perspective. I might for brevity's sake call the errors so resulting errors of superposition. The significance of this concept will, I hope, come to have still greater validity in psychoanalysis. [This error of superposition C. G. Jung attempts to unmask, when he writes: "As libido has a forward tendency, so in a way, incest is that which tends backward into childhood. It is not incest for the child, and only for the adult, who possesses a well constituted sexuality, does this regressive tendency become incest in that he is no longer a child, but has a sexuality that really no longer can suffer a regressive application." Jung, Psychology of the Unconscious.) It may moreover be remarked that Freud also is careful not to take the incest disclosed by psychoanalysis in too physical a sense.] This error of superposition is found not only in the view backward but in the forward view. So what I, as interpreter of mystical symbolism, may say about the possible development of the soul will be affected by this error of superposition. It is not in my power to correct it. In spite of everything, the treatment of symbolism from the two points of view must be superior to the onesided treatment; in order to approximate a fundamental comprehension, which to be sure remains an ideal, the different aspects must be combined and in order to make this clear I have added a synthetic treatment to the analytic part of my work.

Looking back through the elementary types, we

see the infantile images together with those non-
moral origins that psychoanalysis discovers in us;
looking forward we notice thoughts directed to certain goals that will be mentioned later.   The elementary types themselves thanks to intro-determination represent however a collection of our spiritual
powers, which we have first formed and exercised
at the time that the images arose, and which are in
their nature closely related to these images, indeed
completely united with them as a result of the errors
of superposition — this collection of powers, I say,
accompanies us through our entire life and is that
from which are taken the powers that will be required for future development.   The objects or applications change, the powers remain almost the
same.   The symbolism of the material categories
which depends on external things changes with them;
but the symbolism of the functional categories, which
reflects these powers remains constant.   The types
with their intro-determination belong to the functional categories; and so they picture the constant
characters.

That experience to which the suggestions of symbolism (brought to verbal expression by means of
introversion) point as to a possible spiritual development, corresponds to a religious ideal; when intensively lived out this development is called mysticism.   [We can define mysticism as that religious
state which struggles by the shortest way towards
the accomplishment of the end of religion, the union
with the divinity; or as an intensive cultivation of

oneself in order to experience this union.]   It presents itself if instead of looking backward we gaze
forward from our elementary types to the beyond.
But let us not forget that we can regard mysticism
only as the most extreme, and therefore psychically
the most internal, unfolding of the religious life,
as the ideal which is hardly to be attained, although
I consider that much is possible in this direction.
If my later examination carries us right into the
heart of mysticism, without making the standpoint
clear every time, we now know what restrictions we
must be prepared for.

     If I take the view that those powers, whose images
(generally veiled in symbolism) are the elementary
types, do not change, I do not intend to imply that
it is not possible to sublimate them.   With the increasing education of man they support a sublimation
of the human race which yet shows in recognizable
form the fundamental nature of the powers.   One
of the most important types, in which this transformation process is consummated and which refines
the impulse and yet allows some of its character to
remain, is the type mother, i.e., incest.   Among religious symbols we find countless incest images but
that the narrow concept of incest is no longer suited
to their psychological basis (revealed through analysis) has been, among psychoanalysts, quite clearly
recognized by Jung.   Therefore in the case of every
symbolism tending to ethical development, the anagogic point of view must be considered, and most
of all in religious symbolism.   The impulse cor-

responding to the religious incest symbols is pre-
eminently to be conceived in the trend toward introversion and rebirth which will be treated of later.
[Vid. note C, at the end of the volume.]

I have just used the expression " sublimation."
This Freudian term and concept is found in an exactly similar significance in the hermetic writers.
In the receptacle where the mystical work of education is performed, i.e., in man, substances are sublimated; in psychological terms this means that impulses are to be refined and brought from their
baseness to a higher level. Freud makes it clear
that the libido, particularly the unsocial sexual libido,
is in favorable circumstances sublimated, i.e., changed
into a socially available impelling power. This happens in the evolution of the human race and is re-
capitulated in the education of the individual.

I take it for granted that the fundamental character of the elementary psychic powers in which the
sublimation is consummated is the more recognizable
the less the process of sublimation is extended in
time. In mysticism, e.g., the fundamental character
penetrates the primal motive because the latter
wishes to lead the relatively slightly sublimated impulses by a shortened process to the farthest goal
of sublimation. Mysticism undertakes to accomplish in individuals a work that otherwise would
take many generations. What I said therefore
about the unchangeability of the fundamental powers or their primal motive, is wholly true of its fate
in mystical development.

The Mohammedan mystic Arabi (1165–1240) writes, " Love as such, in its individual life, is the same for sensuous and spiritual, therefore equally for every Arab (of an allegory) and for me, but the objects of love are different. They loved sensible phenomena while I, the mystic, love the most intimate existence." (Horten, Myst. Texte, p. 12.)

The religious-mystical applications of the fundamental powers represented by the types, in the sense of a sublimation, does not manifest therefore in contrast to their retrospective form (titanic, purposeless form) an essentially foreign nature; the important novelty in them is that they no longer are used egotistically but have acquired a content that is ethically valuable, to which the intro-determination was an aid. This determination, whose external aspects we have noticed in the types or symbols, is only the visible expression of a far more important actual intro-determination whose accomplishment lies in an amplification of personality, and will later be considered in detail.

In the psychoanalytic consideration of the alchemistic parable it would appear that only the titanic impulses were realized there, e.g., to have the mother as a lover and to kill the father. Now it corresponds to a really significant intro-determination when we hear that in the alchemistic work the father is the same as the son, and when we understand that the father is a state, or psychic potentiality, of the " son," whom the latter in himself, has to conquer, exactly in the same manner as Lea in the lecomantic study strove to put off the old man.

The alchemist Rulandus (Lex., p. 24) quotes the " Turba ": " Take the white tree, build him a round, dark, dew-encircled house, and set in it a hundred year-old man and close it so that no wind or dust can get to him (introversion) ; then leave him there eight days. I tell you that that man will not cease to eat of the fruit of that tree till he becomes a youth. O what a wonderful nature, for here is the father become son and born again." Ibid: " The stone [that is in the anagogic sense, man] is at first the senex, afterwards young, so it is said filius interficit patrem; the father must die, the son be born, die with each other and be renewed with each other."

We must proceed similarly if we wish to interpret the parable anagogically.

What I have already taken from the anagogic fairy tale interpretation as a symbol of introversion shows, of course, also the character of intro-determination.

As for the nature of the relatively unchangeable spiritual tendencies represented by the elementary types [That can also be called in mythological study primal motives] a simple examination of the essentials without any psychological hair splitting, brings us at once to an elementary scheme that will help us to understand the changes (intro-determination) that take place in accordance with the elementary types. We need here only to examine the simplest

reactions of the individual, necessarily produced by
rubbing up again the external world; reactions which
become persistent forms of experience that are approximately as self-evident as the libido itself. The
degree of egoism which is active in the elementary
tendencies must, according to the experience of psychoanalysis, be considered very great. For this purpose I have selected in what follows an excessively
egotistical expression for the " titanic " aspect, the
retrospective form, of the tendencies; and this same
excessive expression which would seem to be rather
objectionable when applied to the basis of a religious
development, enables us, thanks to the principle of
intro-determination, to understand this development.

Starting from the libido in the most general sense
we arrive first of all at the two phenomena, the
agreeableness and the disagreeableness, from which
results at once, acceptance and aversion. Obstacles
may aggravate both activities, so that acceptance
becomes robbery and aversion becomes annihilation.
These possibilities can to be sure only become acts
in so far as they prove practically feasible. In all
cases they are present in the psyche, and in this crude
primal form play no small part in the soul of the
child. It is indeed only a blind sentimentality that
can raise the child to an angelic status, from which
it is as far removed as from its opposite. We
should be careful not to regard the crude form of the
impulse as crude in the sense of an educated humanity, which must see in the crudeness something morally inferior. In robbery and annihilation there

exists on the primitive or childish level hardly the slightest germ of badness. There is much to be said about the psychology and morality of the child. I cannot, however, enter very deeply into this broad topic, interesting though it is.

The primal tendencies, when directed toward the persons in the environment, produce certain typical phenomena. I can unfortunately describe them only with expressions which, if the cultured man uses them, evoke the idea of crime. An ethically colorless language should be made available for these things. [The dream and the myth have found for them the language of symbolism.] The opposition of a fellow man against the working out of an impulse arouses a tendency to overcome this man, to get him out of the way, to kill him. The type of the obstructing man is always the instructor (father, eventually mother). That he is at the same time a doer of good is less appreciated because the psychical apparatus takes the satisfaction of desires as the natural thing, which does not excite its energy nearly as much as does a hindrance to its satisfaction. [Recognition of a good deed, thankfulness, etc., regularly presuppose sublimation; they do not belong to the titanic aspect. A form of appreciation of this kindness however comes to mind. Towards the mother there occurs on the part of the child, though it has been completely overlooked for a long time, very early and gradually increasing, a sexually-toned feeling, although the manifestations of this feeling are very dim and at times may com-

pletely disappear.   In this " love " is contained a
germ of desire, of erotic appropriation-to-self.   Any
woman in the environment and especially the mother
must needs supply the ideal of the desired woman.
In so far as the father is perceived as an obstacle
to the love towards the mother he must, in the
elementary tendency, be killed to remove the obstacle, and there arises the murder impulse belonging
to the Œdipus Complex.   [The child has no clear
idea of death.   It is only a matter of wishing to
have some one out of the way.   If this primal motive appears to us subsequently as a " killing," it is
again only because of the error of superposition, just
as in the later mentioned " rape."]   In so far
as the mother herself does not meet the desired
tenderness or in refusing, acts as a corrective agent,
while carrying on the education, she, too, becomes
an obstacle, a personality contrasting with the
" dear " mother, a contrast which plunges the psyche
in anxiety and bitterness.   Anxiety comes principally
from the conflict of psychical tendencies, which result from the same person being both loved and
hated.   The correlative to the denying action of
the mother is to commit rape on her.   Another cause
of the attraction towards the mother besides the
erotically toned one, is the desire for her care, called
forth by the hardships encountered elsewhere in the
world.   It is an indolence opposed to the duties of
life.   The propensity towards ease is psychologically
a very important factor.   The home is in general
the place of protection; the characteristic embodi-

ment of this is preëminently the mother.   We speak
of maternal solicitude but less of paternal solicitude.
I have noted the solicitous mother type in the story
of the three feathers, where the mother toad be-
stows the gifts from the big box.   In so far as the
solicitous person refuses the requests made of her
and for reasons of necessity thrusts the child out
into the world, or in so far as any other obstacles
(demands of life) stand in the way of the gratification of the lazy, " feed me " state of mind, like
the angel with the flaming sword before the entrance
to paradise, so far the obstructing power appears
as the type of the " terrible " mother, a picture
whose terribleness is yet intensified by the working
of the incest conflict.   In this aspect therefore the
otherwise beloved mother is a hostile personality.

To the process of education on the part of the
parents, felt as pedantry by the child, or to otherwise misunderstood action, he opposes a well known
defiance, and there results, as also from the attempt
to change in general the rough path of life, the
hopeful attempt to get a creative " improvement,"
which I have already discussed.   The wish to die
sometimes occurs.   Further the obstacles that stand
in the way of the full erotic life in the external
world, in so far as they are insuperable or are not
overcome on account of laziness, lead to auto-
erotism.   (That this is found even in early childhood is for the mechanism of the impulses, a side-
issue.   The scheme just given is not to be regarded
as a historical or chronological development, but

the tendencies are quite as intimately connected with each other as with the acquisition of the psychical restraints that are not generally brought to view; in separating them we commit something like an error.)

We have considered the following main forces: 1. Removal of obstacles. 2. Desire for the solicitude of the parents. 3. Desire for the pleasurable [especially of the woman]. 4. Auto-erotism. 5–6. Improvement and re-creation. 7. Death wish. The following scheme shows the retrograde (titanic) as well as the anagogic aspect of these powers, which later corresponds to an intro-determination of the types, and a species of sublimation of impulses.

| Retrograde Aspect. | Anagogic Aspect. |
| --- | --- |
| 1. Killing of the father. | Killing of the old Adam. |
| 2. Desire for the mother (laziness). | Introversion. |
| 3. Incest. | Love towards an Ideal. |
| 4. Auto-erotism | Siddhi.[1] |
| 5. Copulation with the mother. | Spiritual regeneration. |
| 6. Improvement. | Re-creation.[1] |
| 7. Death wish. | Attainment of the ideal. |

We need not scent anything extraordinary behind these intro-determinations, as the scheme is here indeed only roughly sketched; they take place in each and every one of us, otherwise we should be mere beasts. Only they do not in every one of us rise to the intensity of the mystical life.

[1] Explained later.[1]

A more careful inquiry into the mechanism of the psychic powers in the development of mysticism, would show in greater detail how everything that happens is utilized toward intro-determination in the process of education. It would be interesting as an example to discover the application of the special senses to introversion and ascertain the fate of the sense qualities. It is quite remarkable what a prominent rôle tastes and smells often play in descriptions given by persons who have followed the path of mysticism. I mention the odor of sanctity and its opposite in the devilish, evil odor. The experimenter in magic Staudenmaier, who will be mentioned later, has established in his own case the coördination of his partial souls (personifications, autonomous complexes) to definite bodily functions and to definite organs. Certain evil, partial souls, which appear to him in hallucinations as diabolical goat faces, were connected with the function of certain parts of the lower intestine.

Mysticism stimulates a much more powerful sublimation of impulses than the conventional education of men. So it is not strange if intro-determination does not accomplish its desires quickly but remains fragmentary. In such unfortunate or fragmentary cases, the inward-determined powers show more than mere traces of their less refined past. The heroes of such miscarried mysticism appear as rather extraordinary saints. So, for instance, Count von Zinzendorf's warm love of the Savior has so much of the sensual flavor, with furthermore such decided perversities, that the outpourings of his rapture are positively laughable. Thus the pious man indulges his phantasy with a marked predilection for voluptuousness in the "Seitenhölchen" (Wound in the Side) in Jesus' body and with an unmistakable identification of this "cleft" with the vulva.

Examples of the poetical creations of Zinzendorf and his faithful followers are given:

> So ever-sideways-squinting
> So side-homesickness-feeling;
> So lambs-hearts-grave-through crawling,
> So lambs-sweat-trace-smelling.
>
> .   .   .   .   .   .   .   .
>
> So Jesus sweat-drop-yielding,
> With love's fever trembling
> Like the child full of spirit.
> So corpse-air-imbibing,
> So wound-wet-emitting,
> So grave-fume-sniffing.
>
> .   .   .   .   .   .   .   .
>
> So martyr-lamb's heart-like,
> So Jesus-boy-like,
> So Mary Magdelene-like being,
> Childlike, virgin-like, conjugal
> Will the lamb keep us
> Close to the kiss of his clefts.
>
> .   .   .   .   .   .   .   .
>
> With us Cross people
> The closet of the side often is worth
> The whole little lamb.
> Ye poor sinners.

But deep, but deep within,
Yes deep, right deep within,
And whoever will be blessed
He wishes himself within
Into the dear rendezvous
Of all the darlings.
Ravishing little lamb.
I, poor little thing, I kiss the ring
On thy little finger,
Thou wound of the spear
Hold thy little mouth near,
It must be kissed.
Lamb, say nothing to me in there
For this precious minute
Thou art mine only."

On this curiosity compare the psychological explanations of Pfister.    (Frommigkeit G. Ludw. v. Zinzendorf.)

Returning to the previously mentioned " spiritual powers " I should mention that alchemy also attempts to include in a short schema the inventory of powers available for the Great Work.    It uses different symbols for this purpose; one of the most frequent is the seven metals or planets.    Whether I say with the astrologers that the soul (not the celestial spirit, which is derived from God) flowing in from the seven planets upon man, is therefore composed of their seven influences, or if with the alchemists I speak of the seven metals, which come together in the microcosm, it is of course quite the same, but expressed in another closely related symbol.    The metals are, as we know, incomplete and have to be

" improved " or " made complete." That means
we must sublimate our impulses.

" From the highest to the lowest everything rises
by intermediate steps on the infinite ladder, in such
manner that those pictures and images, as outgrowths
of the divine mind, through subordinate divinities
and demigods impart their gifts and emanations to
men. The highest of these are: Spirit of inquiry,
power of ruling and mastering self, a brave heart,
clearness of perception, ardent affection, acuteness
in the art of exposition, and fruitful creative power.
The efficient forces of all these God has above all
and originally in himself. From him they have
received the seven spirits and divinities, which move
and rule the seven planets, and are called angels,
so that each has received his own, distinct from the
rest. They share them again among the seven
orders of demons subordinated to them, one under
each. And these finally transmit them to men.
(Adamah Booz. Sieb. Grunds., p. 9 ff.)

In this enumeration the fundamental powers,
whose partition varies exceedingly, already show a
certain measure of intro-determination. If we wish
to contrast their titanic with their anagogic aspect,
we get approximately the following scheme, to which
I add the familiar astrological characters of the
seven planets.

| | |
|---|---|
| Destroying (castration). | ♄ Introversion. |
| Mastery. | ♃ Mastery of oneself. |
| Love of combat. | ♂ Warring against oneself. |
| Libido. | ☉ Sublimated libido. |

| Sexual life, incest. | ♀ Regeneration. |
| Hypercriticism, fussing. | ☿ Knowledge. |
| Joy in change; Improvement. | ☽ Changing oneself. |

[Freud is of the opinion that the original inquisitive-
ness about the sexual secret is abnormally transformed into morbid over subtlety; and yet can still
furnish an impulsive power for legitimate thirst for
knowledge.]

Beside the partition of the fundamental powers
according to the favorite number seven, there are
to be sure in alchemy still other schemata with other
symbols. We must furthermore continually keep in
mind that the symbols in alchemy are used in many
senses.

In so far as the Constellations, as is often to be
understood in the hermetic art, are fundamental
psychic powers, it sounds just like psychoanalysis
when Paracelsus expresses the view that in sleep
the " sidereal " body is in unobstructed operation,
soars up to its fathers and has converse with the
stars.

With regard to intro-determination I must refer
to my observations in the following sections on the
extension of personality. It is an important fact
that those external obstructions which oppose the
unrestrained unfolding of the titanic impulses are
gradually taken up as constraints into the psyche,
which adopts those external laws, that would make
life practicable. In so far as deep conflicts do not
hinder it, there arises by the operation of these
laws a corresponding influence upon the propensities.

Habit, however, can learn to carry a heavy yoke with love, even to make it the condition of life. I have just made the restriction: if conflicts do not hinder it; now usually these exist, even for the mystics; and the " Work " is above all directed toward their overcoming. For the annihilation of the opposition, the weapons aimed outward in the " titanic " phase must be turned inward; there and not outside of us is the conflict. [Here we see the actual intro-determination briefly mentioned above.]

### B. EFFECTS OF INTROVERSION

Introversion is no child's play. It leads to abysses, by which we may be swallowed up past recall. Whoever submits to introversion arrives at a point where two ways part; and there he must come to a decision, than which a more difficult one cannot be conceived. The symbol of the abyss, of the parting of the ways, both were clearly contained in our parable. The occurrence of the similar motive in myths and fairy tales is familiar. The danger is obvious in that the hero generally makes an apparently quite trivial mistake and then must make extraordinary efforts to save himself from the effects of these few trivial errors. One more wrong step and all would have been lost.

Introversion accordingly presents two possibilities, either to gain what the mystic work seeks, or to lose oneself.

In introversion the libido sinks into " its own depths " (a figure that Nietzsche likes to use), and finds there below in the shadows of the unconscious, the equivalent for the world above which it has left, namely the world of phantasy and memories, of which the strongest and most influential are the early infantile memory images. It is the child's world, the paradise of early childhood, from which a rigorous law has separated us. In this subterranean realm slumber sweet domestic feelings and the infinite hopes of all " becoming." Yet as Mephistopheles says, " The peril is great." This depth is seducing: it is the " mother " and — death. If the libido remains suspended in the wonder realm of the inner world the man has become but a shadow for the world above. He is as good as dead or mortally ill; if the libido succeeds however in tearing itself loose again and of pressing on to the world above, then a miracle is revealed; this subterranean journey has become a fountain of youth for it, and from its apparent death there arises a new productiveness. This train of thought is very beautifully contained in an Indian myth: Once on a time Vishnu absorbed in rapture (introversion) bore in this sleep Brahma, who enthroned on a lotus flower, arose from Vishnu's navel and was carrying the Vedas, eagerly reading them. (Birth of creative thought from introversion.) Because of Vishnu's rapture, however, a monstrous flood overcame the world (swallowing up through introversion, symbolizing the danger of entering into the mother of death). A demon profiting by the danger, stole the vedas from Brahma and hid them in the deep. (Swallowing of the libido.) Brahma wakes Vishnu and he, changing into a fish, dived into the flood, battled with the demon (dragon fight), conquered him and brought the vedas up again. (Prize attained with difficulty.) (Cf. Jung, Psychology of the Unconscious.)

The marvel of the invigoration that can be attained in the successful issue of introversion is comparable to the effect that Antæus felt on touching his mother, the earth. The mother of men, to whom introversion carries us, is the spirit of the race, and from it flows gigantic strength. " This occasional retiring into oneself, which means a return to an infantile relation to the parent images, appears within certain limits to have a favorable effect upon the condition of the individual." Of this mine of power Stekel (Nerv. Angst., p. 375) writes: " When mankind desires to create something big, it must reach down deep into the reservoir of its past."

I wish now to quote a mystic philosopher. J. B. von Helmont (1577–1644) writes: " That magic power of man which is operative outside of him lies, as it were, hidden in the inner life of mankind. It sleeps and rules absolutely without being wakened, yet daily as if in a drunken stupor within us. . . . Therefore we should pray to God, who can be honored only in the spirit, that is, in the inmost soul of man. Hence I say the art of the Cabala requires of the soul that magic yet natural power shall, as it were, after sleep has been driven away, be placed in the keeping of the soul. This magic power has gone to sleep in us through sin and has to be awakened again. This happens either through the illumination of the Holy Ghost or a man himself can by the art of the Cabala produce this power of awakening himself at will. Such are called makers of gold [nota bene!] whose leader (rector) is, however, the spirit of God. . . . When God created the soul of man he imparted to it fundamental and primal knowledge. The soul is the mirror of the universe and is related to all Being. It is illumined by an inner light, but the storm of the passions, the multiplicity of sensuous impressions, and other distractions darken this light, whose beams are spread abroad only, if it burns alone and if all in us is in harmony and peace. If we know how to separate ourselves from all external influences and are willing to be led by this inner light, we shall find pure and true knowledge in us. In this state of concentration the soul distinguishes all objects to which it directs its attention. It can unite with them, penetrate their nature, and can itself reach God and in him know the most important truths." (Ennemoser, Gesch. d. Mag., pp. 906, 914.)

Staudenmaier, who has experimented on himself magically to a great extent and has set down his experiences recently in the interesting book, " Die Magie als experimentelle Naturwissenschaft," thinks he has observed that through the exercise that he carries on, and which produces an intense introversion, psychophysical energies are set free that make him capable of greater efficiency. Specifically, an actual drawing upon the nerve centers unused in the conscious function of the normal man of to-day would be available for intellectual work, etc. So, as it were, a treasure can be gained (by practices having a significant introversion character), a treasure which permits an increased thinking and feeling activity. If Staudenmaier, even in the critical examination of his anomalous functions, can be influenced by them, it would be a great mistake to put them aside simply as "pathological."

Ennemoser says of the danger of introversion (l. c., p. 175): "Now where in men of impure heart, through the destructive natural powers and evil spiritual relations, the deepest transcendental powers are aroused, dark powers may very easily seize the roots of feeling and reveal moral abysses, which the man fixed in the limits of time hardly suspects and from which human nature recoils. Such an illicit ecstasy and evil inspiration is at least recognized in the religious teachings of the Jews and Christians, and the seers of God describe it as an agreement with hell (Isaiah XXVIII, 15)."

Whence comes the danger? It comes from the powerful attraction for us of that world which is opened to us through introversion. We descend there to whet our arms for fresh battles, but we lay them down; for we feel ourselves embraced by soft caressing arms that invite us to linger, to dream enchanting dreams. This fact coincides in large part with the previously mentioned tendency toward comfort, which is unwilling to forego childhood and a mother's careful hands.    Introversion is an excellent road to lazy phantasying in the regressive direction.

Among psychopathologists Jung especially has of late strongly insisted upon the dangerous rôle of indolence.   According to him the libido possesses a monstrous laziness which is unwilling to let go of any object of the past, but would prefer to retain it forever.   Laziness is actually a passion, as La Rochefoucauld brilliantly remarks:   " Of all the passions the least understood by us is laziness; it is the most indefatigable and the most malign of them all, although its outrages are imperceptible."   " It is the perilous passion affecting the primitive man more than all others, which appears behind the suspicious mask of the incest symbols, from which the fear of incest has driven us away, and which above all is to be vanquished under the guise of the ' dreaded mother.'   [Vide, Note D.   To avoid a wrong conception of this quotation it must be noted that laziness is, of course, not to be regarded as the only foundation of incest symbolism.]   She is the mother of infinite evils, not the least of them being the neurotic maladies.   For especially from the vapor of remaining libido residues, those damaging evils of phantasy develop, which so enshroud reality that adaptation becomes well nigh impossible." (Jung, Psychology of the Unconscious.)

That the indolent shrinking back from the difficulties of life is indicated so frequently in psychology and in mythology by the symbol of the mother

is not surprising, but I should yet like to offer for a
forceful illustration an episode from the war of
Cyrus against Astyages which I find recorded in
Dulaure-Krauss-Reiskel (Zeugg., p. 85.) After
Astyages had aroused his troops, he hurled himself
with fiery zeal at the army of the Persians, which
was taken unawares and retreated. Their mothers
and their wives came to them and begged them to attack again. On seeing them irresolute the women
unclothed themselves before them, pointed to their
bosoms and asked them whether they would flee to
the bosoms of their mothers or their wives. This
reproachful sight decided them to turn about and
they remained victorious.

On the origin of the mythological and psychological symbol of the dreaded mother: " Still there
appears to reside in man a deep resentment, because
a brutal law once separated him from an impulsive
indulgence and from the great beauty of the animal
nature so harmonious with itself. This separation
is clearly shown in the prohibition of incest and its
corollaries (marriage laws). Hence pain and in-
dignation are directed toward the mother as if she
were to blame for the domestication of the sons of
men. In order not to be conscious of his desire for
incest (his regressive impulse toward animal nature)
the son lays the entire blame on the mother, whence
results the image of the ' dreaded mother.'
' Mother ' becomes a specter of anxiety to him, a
nightmare." (Jung, Psychology of the Unconscious.)

The snake is to be regarded as a mythological symbol (frequent also in dream life) for the libido that introverts itself and enters the perilous interdicted precinct of the incest wish (or even only the life shirking tendency) ; and especially (though not always valid) is this conception in place, if the snake appears as a terrifying animal (representative of the dreaded mother). So also the dragon is equivalent to the snake, and it can, of course, be replaced by other monsters. The phallic significance of the snake is, of course, familiar enough; the snake as a poisonous terrible animal indicates, however, a special phallus, a libido burdened with anxiety. Jung, who has copious material with which to treat this symbolism, calls the snake really a " negative phallus," the phallus forbidden with respect to the mother, etc. I would recall that alchemy, too, has the symbol of the snake or the dragon, and used in a way that reënforces the preceding conception. It is there connected with the symbols of introversion and appears as " poisonous." The anxiety serpent is the " guardian of the threshold " of the occultists; it is the treasure guarding dragon of the myth. In mystic work the serpent must be overcome; we must settle with the conflict which is the serpent's soul.

Also the mystic yoga manuals of the Hindus know the symbol of the serpent, which the introverting individual has to waken and to overcome, whereupon he comes into possession of valuable powers. These serpents [kundalini] are considered by the Yogi

mystics as an obstacle existing in the human body
that obstructs certain veins or nerves (the anatomy
of the Hindu philosophers is rather loose here), and
by this means, if they are freed, the breath of life
(prāna) sends wondrous powers through the body.
The main path in the body which must be freed for
the increased life-energies is generally described as
the susumna, (As far as I know, it is not yet cleared
up whether the aorta abdominalis or the spine has
furnished the anatomical basis for the idea of the
central canal.) and is the middle way between two
other opposed canals of the breath, which are called
pingala, the right, and ida, the left. (Here, too,
note by the way, appears the comparison of opposites.) I quote now several passages on the kundalini and its significance at the beginning of the
mystical work.

" As Ananta, the Lord of Serpents, supports this
whole universe with its mountains and its forests,
so kundalini is the main support of all the yoga practices. When kundalini is sleeping it is aroused by
the favor of the guru [spiritual teacher], then all
of the lotuses [lotus here stands for nerve center]
and granthis [swallowings, nerve plexus?] are
pierced. Then prana goes through the royal road,
susumna. Then the mind remains suspended and
the yogi cheats death. . . . So the yogi should carefully practice the various mudras [exercises] to
rouse the great goddess [kundalini] who sleeps
closing the mouth of susumna." (Hatha Yoga
Prad., III, 1–5.) " As one forces open a door

[door symbolism] with a key [the " Diederich " of
the wanderer in the parable] so the yogi should
force open the door of moksa [deliverance] by the
kundalini. The Paramesoari [great goddess]
sleeps, closing with her mouth the hole through
which one should go to the brahmarandhra [the
opening or place in the head through which the divine spirit, the Brahma or the Atman, gets into the
body; the anatomical basis for this naïve idea may
have been furnished by one of the sutures of the
skull, possibly the sutura frontalis; the brahmarandhra is probably the goal of the breath that
passes through the susumna that is becoming free.]
where there is no pain or misery. The kundalini
sleeps above the kanda. [The kanda, for which
we can hardly find a corresponding organ, is to be
found between the penis and the navel.] It gives
mukti to the yogis and bondage to the fools. [See
later the results of introspection.] He who knows
her, knows yoga. The kundalini is described as
being coiled like a serpent. He who causes that
sakti [probably, power] to move . . . is freed without doubt. Between the Ganges and the Yamuna
[two rivers of India, which are frequently used symbolically, probably for the right and the left stream
of the breath of life, ingala and ida, cf. what fol-
lows] there sits the young widow [an interesting
characterization of the kundalini] inspiring pity.
He should despoil her forcibly, for it leads one to
the supreme seat of Vishnu. Ida is the sacred
Ganges and pingala the Yamuna. Between ida and

pingala sits the young widow kundalini. You should awake the sleeping serpent [kundalini] by taking hold of its tail. That sakti, leaving off sleep, goes up forcibly." (Hatha-Yoga, Prad., III, 105–111.) Ram Prasad ("Nature's Finer Forces," p. 189) writes about the kundalini: "This power sleeps in the developed organism. It is that power which draws in gross matter from the mother organism through the umbilical cord and distributes it to the different places, where the seminal prana gives it form. When the child separates from the mother the power goes to sleep." Here the kundalini sakti appears clearly in connection with the mother. Siva is the god [father image] most peculiar to the yogis. The wife of Siva, however, is called Kundalini.

Mythologically expressed, introversion proceeds well if the hero defeats the dragon. If this does not happen, an unsuccessful issue is the result; the man loses himself. In my opinion this losing of self is possible in two ways, one active, the other passive. In all there would then be three terminations of introversion. The good conclusion is the entrance into the true mystical work, briefly, mysticism. The bad conclusions are the active way of magic and the passive one of schizophrenia (introversion psychosis). In the first case there is consummated an inner reunion, in the other two cases a losing of self; in magic one loses oneself in passions, for which one wishes to create satisfaction magically, absolving oneself from the laws of nature; in the case of mental malady the sinking develops into laziness, a spir-

itual death. The three paths followed by the introverting individual correspond roughly to these three
other possibilities of life, work (morality), crime,
suicide. These three possibilities are, of course,
recognized by the hermetic art; it recognizes three
fundamental powers, which can give no other result
psychically. Two of these principles are mutually
opposed (in the unpurified condition of the material). We know them quite well as △ and ▽ , etc.
The third principle lies evenly between the other
two, like the staff of Hermes between the two serpents. So the symbol ☿, as Hermes' staff with the
serpents, precisely unites all three. In this aspect
the three qualities or constituents of matter (prakrti)
may at once be susbtituted for the three fundamental
powers of alchemy according to the Hindu samkbrya
doctrine. Sattva, Rajas, Tamas, are translated (by
Schroeder) by " purity, passion, darkness."

In the Bghavad-Gita it is said of the happiness
that these three grant:

" Where one rests after earnest work and arrives at
    the end of toil,
Fortune, which appears poison at first, finally is like
    nectar.
Such a fate is truly good, procured through cheerful-
    ness of spirit.                    [Sattva.]
Fortune that first shows like nectar, and finally ap-
    pears as poison,
Chaining the senses to the world, belongs to the
    realm of passion.                 [Rajas.]

Fortune that immediately and thereafter strikes the
    soul with delusion,
In sleep, indolence, laziness, such Fortune belongs
    to darkness."                              [Tamas.]

"Passion" and "darkness," Rajas and Tamas,
(in alchemy indicated by △ and ▽, also often by ♂
and ♀) indicate the wrong way, the peril in introversion. They lead to what Gorres (Christl.
Myst.) describes as the "demoniac" mysticism as
opposed to the divine mysticism. All mystic manuals warn us of the wrong way and emphasize often
that we can easily lose the way even where there is
good intention. The evil one knows how, by illusions, to make the false way deceptively like the
right one, so that the righteous man, who is not on
his guard, may get unsuspectingly into the worst en-
tanglements. Careful examination of himself, exact
observation of the effect of the spiritual exercises,
is to be laid to heart by every one. Yet powers
come into play that have their roots in the deepest
darkness of the soul (in the unconscious) and which
are withdrawn from superficial view. [After this
had been written I read a short paper of Dr. Karl
Furtmüller, entitled "Psychoanalyse und Ethik,"
and find there, p. 5, a passage which I reproduce
here on account of its agreement with my position.
I must state at the outset that according to Furtmüller, psychoanalysis is peculiarly qualified to
arouse suspicion against the banal conscience, which
leads self-examination into the realm of the conscious

only, with neglect of the unconscious impulses, which are quite as important for the performance of actions. The passage of interest to us here reads: " There is no lack of intimation that these fundamental facts which place the whole of life in a new perspective, were recognized or suspected even in earlier times. If early Christianity believed that demons could overpower the heart of man in the sense that they assumed the voice of God, and the man believed that, while really doing the devil's work he was doing the work of God, then that sounds like a symbolic representation of the play of the forces that are described above." The play of these forces was indeed known to cultivated religious peoples of all times. As for Christianity, what the author asserts of its beginnings can be accepted as true for a much earlier time. We already know that one of the first works of mysticism consists in the education of the conscience, in a most subtle purification of this judicial inner eye. The claims of the psychoanalyst are there fulfilled to the uttermost.] Instead of many examples I gladly quote a single one, but an instructive exposition by Walter Hitton, a great master of the contemplative life, from his " Scala Perfectionis " as Beaumont (Tract. v. Gust. pub. 1721, pp. 188 ff.) renders it. Thus he writes: " From what I said we can to some extent perceive that visions and revelations, or any kind of spirit in bodily appearance, or in the imagination in sleep or waking, or any other sensation in the bodily senses that are, as it were, spiritually performed, either

through a sound in the ears or taste in the mouth or
smell in the nose, or any other perceptible heat of
fiery quality that warms the breast or any other part
of the body, or any other thing that can be felt by
a bodily sense, even if it is not so refreshing and
agreeable, all this is not contemplation or observation; but in respect of the spiritual virtues, and those
of celestial perception and love towards God, which
accompany true contemplation, only evil secondary
matters, even if they appear to be laudable and good.
All such kinds of sensation may be good if produced by a good angel, but may, however, proceed
in a deceptive manner, from the impositions of a
bad angel, if he disguises himself as an angel of light.
For the devil can imitate in bodily sensations exactly
the same things that a good angel can accomplish.
Indeed, just as the good angels come with light, so
can the devil do also. And just as he can fabricate
this in things that appear to the eyes, so he can bring
it to pass in the other senses. The man who has
perceived both can best say which is good and which
is evil. But whoever knows neither or only one, can
very easily be deceived."

Externally, in the sense quality, they are all similar, but internally they are very different. And
therefore we should not too strongly desire them,
nor lightly maintain that the soul can distinguish between the good and evil by the spirit of difference,
so that it may not be deceived. As St. John says:
" Believe not every spirit, but prove it first whether
it be of God or not." And to know whether the

perception of the bodily sense is good or evil, Hitton
gives the following rule:

" If ye see an unusual light or brilliance with
your bodily eye, or in imagination, or if ye hear any
wonderful supernatural sound with your ears, or if
ye perceive a sudden sweet taste in your mouths or
feel any warmth in your breasts, like fire, or any
form of pleasure in any part of your body, or if ye
see a spirit in a bodily form, as if he were an angel
to fortify or instruct you, or if any such feeling that
you know comes not from you or from a physical
creature, then observe yourselves with great care at
such a time and consider the emotions of your heart
prudently.   For if ye become aware by occasion of
pleasure or satisfaction derived from such perception, that your hearts are drawn away from the contemplation of Jesus Christ and from spiritual exercises: as from prayer, and knowledge of yourselves
and your failings, and from the turning in towards
virtue and spiritual knowledge and perception of
God, with result that your heart and your inclina-
tions, your desire and your repose depend chiefly on
the above mentioned feeling or sight, in that ye
therefore retain them, as if that were a part of the
celestial joy or angelic bliss, and therefore your
thoughts become such that ye neither pray nor can
think of anything else, but must entirely give way
to that, in order to keep it and satisfy yourself with
it, then this sensation is very much to be suspected
of coming from the Enemy; and therefore were it

ever so wonderful and striking, still renounce it and do not consent to accept it. For this is a snare of the Enemy, to lead the soul astray by such bodily sensation or agreeableness of the senses, and to trap it in order to hurl it into spiritual arrogance and false security, which happens if it flatters itself as if it enjoyed celestial bliss and on account of the pleasure it feels were already half in paradise, while it is still in fact at the gate of hell, and therefore through pride and presumptuousness may have fallen into error, heresy, fanaticism and other bodily or spiritual disaster.

" In case, however, that these things do not result in leading away your heart from spiritual exercises, but cause ye to become ever more devout and more ardent in prayer and more wise to cultivate spiritual thoughts; if ye are at first astonished but nevertheless your heart turns back and is awakened to greater longing for virtue and your love toward God and your neighbor increases more and more, and makes you ever meeker in your own eyes; then you may infer from this sign that it is of God and comes from the presence and action of a good angel, and comes from the goodness of God, either for comfort to simple pious souls to increase their trust in and longing for God, and because of such a strengthening to seek more thoroughly for the knowledge and love of God. Or if they are perfect that perceive such a pleasure, it appears to them somewhat like a foretaste and shadow of the transfiguration of the body which it may expect in the celestial bliss." However, I do not know whether such a man can be found on earth.

" He continues: Of this method of distinguishing between the works of the spirits, Saint John (I John IV, 3) speaks in his epistle: ' Every spirit that confesseth not that Jesus Christ is come in the flesh is not of God ' (or as it is translated by Luther: ' Who does not recognize that Jesus Christ is come into the flesh '). This union and connection of Jesus with the human soul is caused by a good will and a zealous striving toward him, which alone desires to possess him and to view him spiritually in his blessedness. The greater this longing the more closely is Jesus united with the soul, and the less the longing, the more loosely is he bound to him. So every spirit or every sensation that diminishes this longing, and draws it away from the steadfast contemplation of Jesus Christ and from sighing and longing like a child for him, this spirit will release Jesus from the soul, and therefore it is not from God but the activity of the Enemy. But if a spirit or a sensation increases this desire, fastens the bonds of love and devotion closer to Jesus, raises the eyes of the soul to spiritual knowledge more and more, and makes the heart ever meeker, this spirit is from God."

In many of the modern theosophic introversion methods, borrowed from the Hindu yoga doctrines, we find the exhortation to attach no importance to the marvels appearing beside the real prize, indeed to regard them as a pernicious by-product. The Hindu doctrine calls them Siddhi. Walter Hitton speaks of them as "inferior subordinate matters." From the description of them it appears that they are phantastic appearances, which partly flatter the wish for power, partly other wishes. [See Note E at the end of this volume.] The Siddhi are qualified to captivate weak minds with their jugglery. Erotic experiences are connected very easily with them because, going over into the regressive phase, they show their "titanic" countenances. I have with some daring, but not without right, just cited the Siddhi as the anagogic equivalent of auto-erotism. The regressive phase, however, appears as soon as one indulges in the gratification of the Siddhi. It is not the Siddhi themselves that are the evil (I regard them indeed as anagogic), but the losing of oneself in them. They can be both divine and diabolic. That depends on one's attitude towards them.

In the result of introversion, the diabolic mysticism is opposed, as we saw, to the divine. The true mysticism is characterized by the extension of personality and the false by the shrinking of personality. We can also say, by an extension or shrinking of the sphere of interest that determines the socially valuable attitude. I say advisedly "sphere of interest," for mysticism in the end will not merely fulfill the social law without love, but it labors for the bringing out of this very love. It is not satisfied with superficially tincturing the substance into gold (i.e.,

among other meanings, to get man to do good externally) ; but it would change the substance completely, make it gold through and through (i.e., to
orient the entire impulse power of man for good, so
that he desires this good with the warmth of love
and therefore finds his good fortune in virtue).
Only the good and not the good fortune is chosen
as the leading star, as I must note in order to avoid
a misconception about the hermetic procedure.
Happiness arises only at a certain point, and seems
to me like a fruit ripened in the meantime.   The
most subtle representatives of this doctrine among
the alchemists are not so far, after all, from the
Kantian ethics.

Alchemistic ethics presupposes that there is an
education, an ennobling of the will.   The person
that wills can learn to encompass infinitely much in
his ego. [Cf. Furtmüller (Psychoanalyse und
Ethik, p. 15) : " The individual can . . . make
the commands of others his own." He quotes
Goethe (Die Geheimnisse) :

> From the law which binds all being
> The man is freed who masters himself.

The poles of shrinking and extension are the following:   The magician and the pathological intro-
versionist contract the sphere of their interest upon
the narrowest egoism.   The mystic expands it immensely, in that he comprises the whole world in
himself.   The person egotistically entering into introversion can preserve his happiness only by a firm

self-enclosure before the ever threatening destruction; the mystic is free. The mystic's fortune consists in the union of his will with the world will or as another formula expresses it, in the union with God. [On the freeing effect of the merging of one's own will into a stronger cf. my essays Jb. ps. F., III, pp. 637 ff., and IV, p. 629.] This fortune is therefore also imperishable (gold). The reader must always bear in mind that the mystic never works on anything but on the problem of mankind in general; only he does so in a form of intensive life, and it may indeed be the case that the powers which introversion furnish him, actually make possible a more dynamic activity and a greater result. For my part I am strongly inclined to believe it.

On the extension of personality, some passages from the Discourses on Divinity in the Bhagavad-Gita:

"Who sees himself in all being and all being in himself,
Whoever exercises himself in devotion and looks at all impartially,
Whoever sees me everywhere, and also sees everything in me,
From him I can never vanish nor he from me." VI, 29f.
"Whoever discovers in all the modes of life the very exalted lord,
Who does not fail when they fail — he who recognizes that, has learned well,
For whosoever recognizes the same lord as the one who dwells in all,
Wounds not the self through the self, and travels so the highest road." XIII 27f.

These passages elucidate the progressive function of the idea of God in the " work." Incidentally, I believe that the devotional doctrines (Yoga) which are theoretically based on the Samkhya philosophy that originated without a God, has for good practical reasons taken the idea of isvara (God) into its system. Concentration requires an elevated impalpable object as an aim. And this object must have the property of being above every reach of the power to grasp and yet apparently to seem attainable. God has furthermore the functions of the bearer of conflicts and hopes. At the beginning of the work indeed the obstructing conflicts still exist. A certain unburdening is accomplished by leaving the conflict to the divinity, and frees the powers that were at first crippled under the pressure of the conflicts. [Cf. Jung's Psychology of the Unconscious, Freud Kl. Schr., II, p. 131.]

" Then throw on me all thy doings, thinking only on the
    highest spirit,
Hoping and desiring nothing, so fight, free from all pain."
    Bh. G. III 30.
" Whose acts without any bias and dedicates all his activity
    to God
Will not be stained with evil [is therefore free from con-
    flicts] as the lotus leaf is not stained by the water."
    V 10.

The idea of the education of the will has, of course, been familiar for a long time to ethical writers, even if it has at times been lost sight of.

Aristotle is convinced that morality arises from

custom and convention. " As we learn swimming
only in water, and music by practice on an instrument, so we become righteous by righteous action
and moderate and courageous by appropriate acts.
From uniform actions enduring habits are formed,
and without a rational activity no one becomes good
. . . being good is an act. Good is never by nature; we become good by a behavior corresponding
to a norm. We possess morality not by nature but
against nature. We have the disposition to attain
it . . . we must completely win it by habit. As
Plato says, in agreement with this, the proper education consists in being so led from youth upward,
as to be glad and sorry about the things over which
we should be glad and sorry. But if by a course of
action in accordance with custom, a definite direction of the will has been secured, then pleasure and
pain are added to the actions that result from the
will and, as it were, as signs, that here a new nature
is established in man." (Jodl. Gesch. d. Eth., I,
pp. 44 ff.) " The energy and the proud confidence
in human power with which Aristotle offers to man
his will and character formation as his own work,
the emphasis with which he has opposed to the quietis-
tic " velle non discitur " (we cannot educate volition
nor learn to will, as later pessimistic opinions have
expressed it axiomatically) with the real indispensa-
bility and at the same time the possibility of the
formation of the will; this contention is admirable
and quite characteristic of the methods of thought
of ancient philosophy at its height." (Jodl., l. c.,

p. 49.)    [Velle non discitur has been popularized
by Schopenhauer.]

In Philo and the related philosophers there appears quite clearly the thought that gained such wide
acceptance later among the Christian ascetics, that
the highest development of moral strength was attainable only through a long continued and gradually
increasing exercise, an ethical gymnastics.    Philo,
moreover, uses the word Askesis to describe what
elsewhere had been described as bodily exercise.
The occidental spiritual exercise corresponds to the
Hindu yoga.

In the domestication of man through countless
generations, social instincts must have been established, which appear as moral dispositions.    I recall
the moral feeling in Shaftesbury.    The social life
of man, for instance, plays with Adam Smith a significant rôle, and yet even with him the moral law
is not something ready from the very beginning, not
an innate imperative, but the peculiar product of
each individual.    The development of conscience receives an interesting treatment by Smith.    There
takes place in us a natural transposition of feelings,
mediated through sympathy, which arouse in each
of us the qualities of the other, and we can say " that
morality in Smith's sense, just as Feuerbach taught
later, is only reflected self-interest, although Smith
himself was quite unwilling to look at sympathy as
an egotistic principle.    By means of a process that
we can almost call a kind of self-deception of the
imagination, we must look at ourselves with the eyes

of others, a very sensible precaution of nature, which thus has created a balance for impulses that otherwise must have operated detrimentally. [Bear in mind what I have said above about intro-determination.] This transposition which sympathy effects we cannot escape; it itself appears when we know that we are protected from the criticism of another by the complete privacy of our own doings. It alone can keep us upright when all about us misunderstand us and judge us falsely. For the actual judgments of another about us form, so to speak, a first court whose findings are continually being corrected by that completely unpartisan and well informed witness who grows up with us and reacts on all our doings." (Jodl., l. c., I, pp. 372 ff.)

The derivation of the moral from selfish impulses by transposition does not resolve ethics into egoism, as Helvetius would have us believe. It is " a caricature of the true state of things to speak of self-interest, when we have in mind magnanimity and beneficence, and to maintain that beneficence is nothing but disguised selfishness, because it produces joy or brings honor to the person that practices it." (L. c., p. 444.)

The ethical evolution which takes place as an extension of personality demands, the more actively it is practiced, the removal of resistances which operate against the expansion of the ego. It cannot be denied that hostile tendencies, which are linked with pusillanimous views, are always on hand and create conflicts. If they were not, the moral task would

be an easy one.    Now as man cannot serve two masters, so in the personal psychical household, the
points of view which have been dethroned, as far
as they will not unite with the newly acquired ones,
must be killed, and ousted from their power.    Most
of all must this process be made effective if the development is taken up intensively in the shape of
introversion.    It must appear also in the symbolism.

Already in the lecanomantic experiments we are
struck by the dying of the figure (old man) that represents the old form of conscience that has been overcome.    It is that part of Lea's psyche that resists
the new, after the manner of old people (father
type).    In order that the new may be suppressed,
it must be immolated; at every step in his evolution
man must give up something; not without sacrifice,
not without renunciation, is the better attained.
The sacrifice must come, of course, before the new
reformed life begins.    The hermetic representations
do not indeed always follow chronological order,
yet the sacrifice is usually placed at the beginning, as
introversion.    In the parable the wanderer kills the
lion, well at the beginning.    He sacrifices something
in so doing.    He kills himself, i.e., a part of himself,
in order to be able to rise renewed (regenerated).
This process is the first mystical death, also called
by the alchemists, putrefaction or the blacks.    This
death is often fused with the symbol of introversion,
because both can appear under the symbol of the
entrance into the mother or earth.    Only by closer

examination can it sometimes be seen which process is chiefly intended.

" And that shalt thou know my son, whoso does not know how to kill, and to bring about a rebirth, to make the spirits revive, to purify, to make bright and clear . . . he as yet knows nothing and will accomplish nothing." (Siebengestirn, p. 21.)

" These are the two serpents sent by Juno (which is the metallic nature) which the strong Hercules (i.e., the wise man in his cradle) has to strangle, i.e., to overpower and kill, in order in the beginning of his work to have them rot, be destroyed and to bear." (Flamel, p. 54.)

Again and again the masters declare that one cannot attain to true progress except by means of the blacks, death and putrefaction.

In the " Clavis philosophiae et alchymiae Fluddanae," of the year 1633, we read: " Know then that it is the duty of spiritual alchemy to mortify and to refine all obscuring prejudice as corruptible and vain, and so break down the tents of darkness and ignorance, so that that imperishable but still beclouded spirit may be free and grow and multiply in us through the help of the fiery spirit, full of grace, which God so kindly moistened, so as to increase it from a grain to a mountain. That is the true alchemy of which I am speaking, that which can multiply in me that rectangular stone, which is the cornerstone of my life and my soul, so that the dead in me shall be awakened anew, and arise from the old

nature that had become corrupted in Adam, as a
new man who is new and living in Christ, and therefore in that rectangular stone. . . ."

To the "sacrifice" of the person introverting,
Jung devotes an entire chapter in his Psychology of
the Unconscious, Chapt. IV.   A brief résumé of it
would show that by the sacrifice is meant the giving
up of the mother, i.e., the disclaiming of all bonds
and limitations that the soul has carried over from
childhood into adulthood.   The victory over the
dragon is equivalent to the sacrificing of the regressive (incestuous) tendency.   After we have sought
the mother through introversion we must escape
from her, enriched by the treasure which we have
gotten.

The sacrifice of a part of ourself (killing of the
dragon, the father, etc.) is, as Jung points out, represented also in mythology by the shooting with
sharp arrows at the symbol of the libido.   The symbol of the libido is generally a sun symbol.   Now
it is particularly noteworthy that the VIII key of
the alchemist Basilius Valentinus (see figure 3, p.
199) shows arrows being shot, which are aimed at
the ☉ (this libido symbol par excellence) that is
aptly used as a "target."   Death is clearly enough
accentuated and correlated with the sinking of the
corns of wheat into the earth.   [John XII, 24, 25,
Verily, verily, I say unto you, Except a corn of wheat
fall into the ground and die, it abideth alone; but
if it die, it bringeth forth much fruit.   He that
loveth his life shall lose it; and he that hateth his

life in this world shall keep it unto life eternal.]
As this rises, so also will the dying mystic rise. The
grave crosses have the form △ (♀); they show that
the interred one is a certain sulphur, the impure sulphur, willfullness. The birds, from which we are
to protect the grain, may in the end be the Siddhi;
they are, in the introversion form of the religious
work, what would otherwise be merely " diversions "
or " dissipations."

The mystical death is the death of egoism (in
Hindu terminology ahamkāra). Jacob Boehme
writes in his book of the true antonement, I, 19:
". . . Although I am not worthy, [Jesus] take me
yet in thy death and let me only in thy death die my
death; still strike thou me in my ancknowledged selfishness to the ground and kill my selfishness by thy
death. . . ." In the Mysterium Magnum, XXXVI,
74, 75: ". . . We exalt not the outspoken word
of the wisdom of God, but only the animal will to
selfishness and egoism which is departed from God,
which honors itself as a false God of its own and
may not believe or trust God (as the Antichrist who
has placed himself in God's stead); and we teach on
the contrary that the man of the Antichrist's image
shall wholly die so that he may be born in Christ of
a new life and will, which new will has power in the
perfect word of nature with divine eyes to see all the
miracles of God, both in nature and creature, in the
perfect wisdom. For as dies the Antichrist in the
soul, so rises Christ from the dead."

In the hermetic book, " Gloria Mundi," it is re-

lated of Adam that he would have been able, if he
had not acted contrary to God, to live 2000 years
in paradise and would then have been taken up into
heaven; but he had drawn on himself death, sickness
and calamity.    Only through the grace of God was
he given a partial knowledge of the powers of
things, of herbs and remedies against manifold in-
firmities.    "When, however, he could no longer
maintain himself by the medicinal art [in paradise]
he sent his son Seth forth to paradise for the tree
of life, which he received, not physically, but spiritually.    Finally he desired the oil of compassion,
whereupon by the angels, at God's command to give
the oil, the promise was given and thereupon the
seed of the oil tree sent, which seed Seth planted
on his return, after his father's death and on his
father's grave, from which grew the wood of the
holy cross, on which our Lord Jesus Christ, through
his passion and death, freed us from death and all
sins; which Lord Christ in his holiest humanity has
become the tree and the wood of life and has brought
to us the fruit of the oil of compassion. . . ."
Adam is the undomesticated man; this ideal must die
to the moral aspirant.

The painful duty of killing a part of self is beautifully expressed in the Bhagavad-Gita, where the
hero, Aryuna, hesitates to fight against his "kin-
dred," to shoot at them — the bow falls from his
hand.

Dying relates to the old realms.    The old laws
expire to make room for the new.    The new life

cancels the old deeds. (Cf. Paul, Rom. VII–VIII.)

Vedanta doctrine: But as to the duty of the scripture canon and perception, both last as long as Samsāra, i.e., until the awakening. If this is attained, perception is annulled, and if you derive thence the objection that thereby the veda is annulled, it must be noted that according to our own doctrine father is not father and the Veda is not the veda. (Deussen, Syst. d. Ved., p. 449.)

Bhagavad-Gita, IV, 37:

" Like fire when it flames and turns all the firewood
to ashes."

So the fire of knowledge burns for you all deeds
to ashes

For several reasons the father image is peculiarly suited to represent what has to be resolved. By the father, the old Adam (totality of inherited instincts) and the strongest imperatives are implanted in the child. The father is also the type of tenacious adherence to the ancestors. Again we meet the antithesis, old generation, new generation, in ourselves after the intro-determination.

The mystical death (sacrifice) is not to be accomplished by mere asceticism, as it were, mechanically; the alchemists warn us carefully against severe remedies. The work is to take a natural course; the work is also, although indeed a consummation of nature, yet not above nature.

> " Nature rejoices in nature
> Nature overcomes nature
> Nature rules nature."

Thus the magician Osthanes is said to have taught. And the Bhagavad-Gita (VI, 5–7) says:

" Let one raise himself by means of self, and not abase self,
Self is his own friend, is also his own enemy.
To him is his self his own friend, who through self conquers self,
Yet if it battle with the external world, then self becomes enemy to self."

In the " Clavis Philosophiae et Alchymiae Fluddanae " (p. 57) we read: " So it is impossible to rise to the supramundane life, in so far as it does not happen by means of nature. From the steps of nature Jacob's ladder is reached and the chain to Jupiter's throne begins on earth."

The idea of self-sacrifice (with dismemberment) appears very prettily in an allegorical vision of the old hermetic philosopher Zosimos, who seems to have copied it, as Reitzenstein notes, from an Egyptian Nekyia. I quote from Hoefer (Hist. Chim., I, pp. 256–259):

" I slept and saw a priest standing before an altar shaped like a cup and with several steps by which to climb to it. [First 15, later 7 steps are mentioned.] And I heard a voice crying aloud, ' I have finished climbing and descending these 15 steps, resplendent with light.' After listening to the priest officiating at the altar I asked him what this resounding voice was whose sound had struck my ear. The priest answered me, saying: ' I am he who is ($\epsilon i \mu i$ $\dot{o}$ $\ddot{\omega} \nu$), the priest of the sanctuary, and I am under the weight of the power that overwhelms me. For at the break of day came a deputy who seized me, killed me with a sword, cut me in pieces; and after flaying the skin from my head, he mixed the bones with the flesh and burned me in the fire to teach me that the spirit is born with the body. That is the power that overwhelms me.' While the priest was saying that, his eyes became as blood, and he vomited all his flesh. I saw him mutilate himself, rend himself with his teeth and fall on the ground. Seized with terror I awoke, and I began to ponder and ask myself if this indeed was the nature and the composition of the water. And I congratulated myself upon having reasoned well [namely in a train of thought preceding the vision]. Soon I slept again and perceived the same altar, and on this altar I saw water boiling with a noise and many men in it. Not finding any one in the neighborhood to explain this phenomenon, I advanced to enjoy the spectacle at the altar. Then I noticed a man with gray hair and thin, who said to me, 'What are you looking at?' 'I am looking,' I answered with surprise, ' at the boiling of the water and the men who are boiling there still alive.' 'The sight you see,' replied he, ' is the beginning, the end and the transmutation ($\mu\epsilon\tau\alpha\beta o\lambda\dot{\eta}$).' I asked him what the transmutation was. 'It is,' he said, ' the place of the operation which is called purification [in the original, topos askeseos], for the people who wish to become virtuous come there and become spirits shunning the body.' And I asked him, ' Are

you also a spirit [pneuma]?' 'I am,' said he, 'a
spirit and the guardian of spirits.' During this conversation and amid the noise of the boiling water
and the cries of the people, I perceived a man of
brass, holding in his hand a book of lead, and I heard
him tell me in a loud voice: 'See, I command all
those who are subjected to punishments to learn from
this book. I command every one to take the book
of lead and to write in it with his hand until his
pharynx is developed, the mouth is opened, and
the eyes have taken their place again.' The act
followed the word, and the master of the house,
present at this ceremony, said to me, 'Stretch your
neck and see what is done.' 'I see,' said I. 'The
brazen man that you see,' said he, 'and who has left
his own flesh, is the priest before the altar. It is
he who has been given the privilege of disposing of
this water.' In going over all this in my imagination I awoke and said to myself, 'What is the cause
of this occurrence? What indeed is it? Is it
not the water white, yellow, boiling, divine?' I
found that I had reasoned well. . . . Finally, to
be brief, build, my friend a temple of a single stone
[monolith] . . . a temple that has neither beginning nor end, and in the interior of which there is
found a spring of purest water, and bright as the
sun. It is with the sword in hand that one must
search and penetrate into it, for the entrance is narrow. It is guarded by a dragon, which has to be
killed and flayed. By putting the flesh and the bones
together you make a pedestal up which you will

climb to reach the temple, where you will find what you are looking for. For the priest, who is the brazen man whom you saw sitting near the spring, changes his nature and is transformed into a man of silver, who can, if you wish, change himself into a man of gold. . . . Do not reveal anything of this to any one else and keep these things for yourself, for silence teaches virtue. It is very fine to understand the transmutation of the four metals, lead, copper, tin, silver, and to know how they change into perfect gold. . . ."

Psychoanalysis, like comparative mythology, makes it probable that the killing or dismemberment of the father figure is equivalent to castration. That has, according to intro-determination, an anagogic, a wider sense, if we compare the organ of generation to the creative power, and a narrower, if we compare it to sexuality. The wider conception does not require immediate interpretation. With regard to the narrower, I observe that the mystical manuals show that the most active power for spiritual education is the sexual libido, which for that reason is partially or entirely withdrawn from its original use. (Rules of chastity.) " Vigor is obtained on the confirmation of continence." (Patanjali, Yoga-Sutra, II, 38.) These instruction books have recognized the great transmutability of the sexual libido. (Cf. ability of sublimation in the alchemistic, as well as in the Freudian terminology.) Naturally the reduction of sexuality had to occur at the beginning of the work in order to furnish that power; hence the castration at the commencement of the process. The killing of the phallic snake amounts, of course, to the same thing. The snake with its tail in its mouth is the cycle of the libido, the always rolling wheel of life, of procreation, which always procreates itself, and of the creation of the world. The same cycle is represented by a god who holds his phallus in his mouth, and so (in accordance with infantile and primitive theory) constantly impregnates himself. The serpent is good and also evil. Whoever breaks through the ring frees himself from the wheel of compulsion, raises himself above good and evil, in order to put in its place later a mystical union [Hieros Gamos].

Regarded from the point of view of knowledge, the formation of types reveals itself as a symbolic presentiment of an anagogic idea, not at first clearly conceivable. For the spirit, what cannot yet be clearly seen (mythological level of knowledge) or can no longer be seen (going to sleep, etc.) is pictured in symbolic form. [More details will be found in my essays, " Phantasie und Mythos," " Ueber die Symbolbildung," and " Zur Symbolbildung " (Jb. ps. F., II, III, IV).] This symbol form is the form of knowledge adapted to the spirit's capacity as it then existed. Not that any mysterious presentiment or prophetic gift of vision must be assumed. The circumstance that man can get ever deeper meaning from his symbols gives them the appearance of being celestial harbingers sent forth by the latest ideas that they express. In a certain

sense, however, the last meaning is implicated in the
first appearance of the typical symbol. It has already been explained by intro-determination how
that was possible. The psyche, whose inventory of
powers is copied symbolically in the elementary
types, knows, even if only darkly at first, the possible
unfolding of the powers. These unfoldings are
originally not actual but potential. [See Note F.]

The more then that the psyche ·is so developed,
that what was originally only a possible presentiment of actuality and that hence tends to come
nearer the merely potential, begins to become actual,
the more symbolism has the value of a " program."
According to Jung, Riklin, etc., the phantasy (dream,
myth-making) can be conceived not only as with
Freud, " as a wish fulfillment, wherein older and
infantile material expresses the wish for something
unsettled, unattained or suppressed, but also as a
mythological first step in the direction of conscious
and adapted thinking and acting, as a program.
. . . Maeder has discussed the teleological functions of the dream and the unconscious. In the
course of an analytic treatment we discover the continuous transformations of the libido symbol in
the dream current, till a form is reached which
serves as an attempt to adapt oneself to actuality.
There are epochs in the history of civilization which
are particularly characterized by a storing of the
libido in the sense that from the reservoir of mythological and religious thought forms, new adaptations
to the real processes and data are made. A signifi-

cant example is the Renaissance, which a study of renaissance literature and a visit to the renaissance cities, e.g., Florence, make evident in a high degree. The analysis of romanticism . . . confirms these processes of development." (Zentralblatt f. Psa., III, p. 114.)

We have here the thought that the " program " is expressed in art, which therefore has prescience in a certain degree of the coming event. Jung (Jb. ps. F., III, pp. 171 ff.) writes: " It is a daily experience in my professional work (an experience whose certainty I must express with all the caution that is required by the complexity of the material) that in certain cases of chronic neuroses, a dream occurs at the time of the onset of the malady or a long time before, frequently of visionary significance, which is indelibly imprinted on the memory and holds a meaning, concealed from the patient, which anticipates the succeeding experiences, i.e., the psychological signficance. Dreams appear to stay spontaneously in memory so long as they suitably outline the psychological situation of the individual."

The more the program is worked out the more the value of the symbolism (whose types can always remain the same in spite of changes in their appearance) changes into that of the functional symbolism in the narrower sense; for the functional symbolism in the restricted sense is that which copies the actual play of forces in the psyche.

To the functional symbolism of actual forces belong, e.g., in large part the faces in my lecanomantic experiments, although they also contain program material; further, in purest form, the previously related autosymbolic vision of the mountains. The progress of a psychoanalytic treatment is, apart from the program connections, generally copied in the dream in correspondence to the momentary psychic status, and therefore actually and functionally. It is quite probable that the progress of the mystical work is represented to the mystic in his phantasying (dreams, visions, etc.) in a symbolic manner. But when one happens upon written phantasy products of the mystics, of course only he who has mystical experiences of his own can venture to say whether a program symbolism or an actually functional symbolism is exhibited. For example, I make no judgment on the degree of actuality in the anagogic symbolism of the parable.

### C. REGENERATION

In the favorable issue of introversion, i.e., when we conquer the dragon, we liberate a valuable treasure, namely, an enormous psychic energy, or, according to the psychoanalytic view, libido, which is applicable to the much desired new creation (as the titanic aspect of which we recognize the " reforming "). The symbolic type, either openly or hiddenly expressed, of the setting free of an active libido, is birth. A libido symbol with the characteristic of active life comes out of a mother symbol. (The former is either explicitly a child or even a food, or it is phallic or animal. Zbl. Psa., III, p. 115.) As

the mystic is author of this, his birth, he has become
his own father.

Introversion (seeking for the uterus or the grave)
is a necessary presupposition of regeneration or
resurrection, and this is a necessary presupposition
of the mystical creation of the new man.    (John III,
1–6) :    "There was a man of the Pharisees named
Nicodemus, a ruler of the Jews.    The same came to
Jesus by night [introversion] and said unto him,
' Rabbi, we know that thou art a teacher come from
God; for no man can do these miracles that thou
doest except God be with him.'    Jesus answered and
said unto him, Verily, verily, I say unto thee, Except
a man be born again, he cannot see the kingdom of
God.    Nicodemus saith unto him, How can a man
be born when he is old?    Can he enter the second
time into his mother's womb and be born?    Jesus
answered, Verily, verily, I say unto thee, Except a
man be born of water and of the Spirit, he cannot
enter into the kingdom of God.    That which is born
of the flesh is flesh, that which is born of the Spirit
is spirit."

Water is one of the most general religious mother
symbols (baptism).    With the earliest alchemists
the brazen man becomes silver, the silver man, gold,
by being dipped in the holy fountain.

A mythological representation of introversion
with its danger and with regeneration was given previously [see Vishnu's adventure].    Detailed examples follow; first the Celtic myth of the birth of
Taliesin.

In olden times there was a man of noble parentage in Peelyn named Tegid Voel. His ancestral country was in the center of the lake of Tegid. His wife was called Ceridwen. Of her he had a son, Morvram ap Tegid, and a daughter, Creirwy, the fairest maiden in the world. These two had another brother, the ugliest of all beings, named Avagddu. Ceridwen, the mother of this ill favored son, well knew that he would have little success in society, although he was endowed with many fine qualities. She determined to prepare a kettle [introversion] for her son, so that on account of his skill in looking into the future [Siddhi] he should find entrance into society. The kettle of water began to boil [cooking of the child in the uterus vessel] and the cooking had to be continued without interruption till one could get three blessed drops from the gifts of the Spirit [treasure]. She set Gwyon, the son of Gwreang of Llanveir, to watch the preparation of the kettle, and appointed a blind man [mutilation or castration] named Morda to keep alight the fire under the kettle, with the command that he should not permit the interruption of the boiling for a year and a day. [Cf. the activity of the wanderer in the parable, Sec. 14 ff.] Meanwhile Ceridwen occupied herself with the stars, watched daily the movement of the planets, and gathered herbs of all varieties that possessed peculiar powers [Siddhis]. Towards the end of the year, while she was still looking for herbs, it happened that three drops of the powerful water flew

out of the kettle and fell on Gwyon's finger.   They
scalded him and he stuck his finger in his mouth.
As the precious drops touched his lips all the events
of the future were opened to his eyes, and he saw
that he must be on his guard against Ceridwen
[dreaded mother].   He rushed home.   The kettle
split into two parts [motive of the tearing apart of
the uterus], for all the water in it except the three
powerful drops were poisonous [danger of introver-
sion], so that it poisoned the chargers of Gwyddno
Garantur, which were drinking out of the gutter
into which the kettle had emptied itself [the flood].
Now Ceridwen came in and saw that her whole
year's work was lost.   She took a pestle and struck
the blind man so hard on the head that one of his
eyes fell out on his cheeks.   " You have unjustly
deformed me," cried Morda; " you see that I am
guiltless.   Your loss is not caused by my blunder."
" Verily," said Ceridwen, " Gwyon the Small it was
that robbed me."   Immediately she pursued him,
but Gwyon saw her from a distance and turned into
a hare and redoubled his speed, but she at once became a hound, forced him to turn around and chased
him towards a river.   He jumped in and became a
fish, but his enemy pursued him quickly in the shape
of an otter, so that he had to assume the form of a
bird and fly up into the air.   But the element gave
him no place of refuge, for the woman became a
falcon, came after him and would have caught him
[forms of anxiety].   Trembling for fear of death,
he saw a heap of smooth wheat on a threshing floor,

fell into the middle of it and turned into a grain of
wheat. But Ceridwen took the shape of a black
hen, flew to the wheat, scratched it asunder, recognized the grain and swallowed it [impregnation, in-
cest]. She became pregnant from it and after being
confined for nine months [regeneration] she found
so lovely a child [improvement] that she could no
longer think of its death [immortality]. She put it
in a boat, covered it with a skin [skin = lanugo of
the fœtus, belongs to the birth motive], and at the
instigation of her husband cast the skiff into the
sea on the 29th of April. At this time the fish weir
of Gwyddno stood between Dyvi and Aberystwyth,
near his own stronghold. It was usual in this weir
every year on the 1st of May to catch fish worth
100 pounds. Gwyddno had an only son, Elphin.
He was very unfortunate in his undertakings, and
so his father thought him born in an evil hour. His
counselors persuaded the father, however, to let his
son draw the weir basket this time, to try whether
good luck would ever be his, and so that he might
yet gain something with which to go forth into the
world. On the next day, the 1st of May, Elphin
examined the weir basket and found nothing, yet as
he went away, he saw the boat covered with the
skin rest on the post of the weir. One of the fisher-
men said to him, " You have never been so unlucky
as you were to-night, but now you have destroyed the
virtue of the weir basket," in which they always
found a hundred pounds' worth on the first of May.
" How so? " asked Elphin. " The boat may easily

contain the worth of the hundred." The skin was lifted and he that opened it saw the forehead of a child and said to Elphin, " See the beaming forehead." ." Beaming forehead, Taliesin, be his name," replied the prince, who took the child in his arms and because of his own misfortune, pitied it. He put it behind him on his charger. Immediately the child composed a song for the consolation and praise of Elphin, and at the same time prophesied to him his future fame. Elphin took the child into the stronghold and showed him to his father, who asked the child whether he was a human being or a spirit. Whereupon he answered in the following song: " I am Elphin's first bard; my native country is the land of the cherubim. The heavenly John called me Merddin [Merlin] and finally, every one, King: Taliesin. I was nine months in the womb of my mother Ceridwen, before which I was the little Gwyon, now I am Taliesin. With my Lord I was in the world above, and fell as Lucifer into the depths of hell. I carried the banner before Alexander. I know the names of the stars from north to south. I was in the circle of Gwdion [Gwydi on] in the Tetragrammaton. I accompanied the Hean into the valley of Hebron. I was in Canaan when Abraham was killed. I was in the court of Dve before Gwdion was born, a companion of Eli and Enoch. I was at the judgment that condemned the Son of God to the cross. I was an overseer at Nimrod's tower building. I was in the ark with Noah. I saw the destruction of Sodom. I was in Africa

before Rome was built. I came hither to the remains of Troy (i.e., to Britain, for the mystical pro-
genitor of the Britons boasted a Trojan parentage).
I was with my Lord in the asses' manger. I com-
forted Moses in the Jordan. I was in the firmament with Mary Magdalene. I was endowed with
spirit by the kettle of Ceridwen. I was a harper at
Lleon in Lochlyn. I suffered hunger for the son of
the maiden. I was in the white mountains in the
court of Cynvelyn in chains and bondage, a year and
a day. I dwelt in the kingdom of the Trinity [Tri-
unity]. It is not known whether my body is flesh
or fish. I was a teacher of the whole world and
remain till the day of judgment on the face of the
earth. [Briefly, Taliesin has the ubiquity of ☿.]
I sat on the shaken chair at Caer Seden [Caer Seden
is probably the unceasingly recurrent cycle of animal
life in the center of the universe.], which continually
rotates between the three elements. Is it not a marvel that it does reflect a single beam?" Gwyddnaw,
astonished at the evolution of the boy, requested another song and received the answer: "Water has
the property of bringing grace; it is profitable to devote one's thoughts aright to God; it is good warmly
to pray to God, because the grace which goes out
from him cannot be thwarted. Thrice have I been
born; I know how one has to meditate. It is sad
that men do not come to seek all the knowledge of
the world, which is collected in my breast, for I know
everything that has been and everything that will
be." (Nork. Myth. d. Volkss., pp. 662 ff.)

The story of Taliesin closely harmonizes with that of Hermes in the Smaragdine tablet. Nork makes some interesting observations, which besides the nature myth interpretation, contains also an allusion to the idea of spiritual regeneration.

I have already mentioned that the uterus symbol is frequently the body cavity of a monster. Just as in the previous myth the hero by introversion gets three marvelous drops, so in the Finnish epic Kalevala, Wäinämöinen learns three magic words in the belly of a monster, his dead ancestor Antero Wipunen. The gigantic size of the body of the being that here and in other myths represents the mother, has an infantile root. The introverting person, as we know, becomes a child. To the child the adults, and of course, the mother, are very large. For the adult, who becomes a child and revives the corresponding images, the mother image may easily become a giant.

Stekel tells (Spr. d. Tr., p. 429) of a patient whose dreams show uterus and regeneration phantasies in concealed form, that he, advised of it by Stekel, mused upon it some minutes and then said, " I must openly confess to you these conscious phantasies. I was 13 years old when I wished to become acquainted with an enormously large giantess, in whose body I might take a walk, and where I could inspect everything. I would then make myself quite comfortable and easy in the red cavern. I also phantasied a swing that was hung 10 m. high in the body of this giantess. There I wanted to swing up

and down joyfully." This patient had carried over
the original proportion of fœtus and mother to his
present size. Now that he was grown up, the body
in which he could move had to be the body of a
giantess.

We shall now not be surprised at the flesh mountain Krun of the mandæan Hibil-Ziwâ saga or similar giant personalities. Hibil-Ziwâ descended into
the world of darkness in order to get the answer to
a question (i.e., once more the treasure in the form
of a marvelous word). He applied in vain to different persons, but always had to go deeper and finally
came to Krûn, from whom he forced the magic word.

The treasure or wonder working name comes
from the depths according to the hermetic cabalistic
conception also. David is supposed to have found
at the digging of the foundation of the temple, the
Eben stijjah, Stone of the Deeps, that unlocked the
fountain of the great deep (I Mos., VII, 11, and
VIII, 12) and on which the Sêm ha-mephoraś, the
outspoken name (of God) was inscribed. This
stone he brought into the holy of holies, and on it
the ark of the covenant was set. Fearless disciples
of wisdom entered at times into the sanctuary and
had learned from the stone the name with its combinations of letters in order to work wonders there-
with.

In cases where the uterus is represented by the
body cavity of a monster the rebirth occurs most
frequently by a spitting forth. Also the breaking
forth by means of tearing apart the uterus occurs,

and in every case it has the significance of a " powerfully tearing of oneself away," the burning of bridges behind one, the final victory over the mother. To the descent into the underworld (introversion) corresponds, as characteristic of the subsequent rebirth, the rising to the light with the released treasure (magic word as above, water of life, as in Ishtar's hell journey, etc.).

A frequently used symbol for the released libido is the light, the sun. Reborn sun figures, in connection with a daily and yearly up and down, are also quite general. That the released libido appears thus may have several reasons. External ones, like the life-imparting properties of the sun, invite comparison. Then the parallel light = consciousness. [Also that higher or other consciousness that is mediated by the mystic religious work; for which expressions like illuminate, etc., are sufficiently significant. On this topic see my essay, Phant. u. Myth. (Jb., II, p. 597).] and also inner reasons, i.e., such as rest upon the actual light and warmth sensations, which occur, as literature and observations show, in persons who are devoted to spiritual training. Here the occasion may be offered to the mystic to utilize for conscious life and action, functions that hitherto had been unconscious. Of the appearance of light in the state of introversion, the histories of saints and ecstatics, and the autobiographies of this kind of men are full. An enormous number of instances might be given. I shall rest content with recalling that Mechthildis von Magdeburg has entitled her revelations: "A flowing Light of my Godhead" ("Ein vliessend Lieht miner Gotheit"), and with adding Jane Leade's words: "If any one asks what is the magic power [sought by the reborn] I answer, 'It is to be compared to a wonderfully powerful inspiration to the soul, to a blood, coloring and penetrating and transmuting the inner life, a concentrating and essentially creative light and fire flame.'"

The Omphalopsychites or Hesychiasts, those monks who dwelt in the Middle Ages on Mount Athos, were given the following instructions by their Abbot Simeon: "Sitting alone in private, note and do what I say. Close thy doors and raise thy spirit from vain and temporal things. Then rest thy beard on the breast and direct the gaze with all thy soul on the middle of the body at the navel. [See Note G.] Contract the air passages so as not to breathe too easily. Endeavor inwardly to find the location of the heart, where all psychic powers reside. At first thou wilt find darkness and inflexible density. When, however, thou perseverest day and night, thou wilt, wonderful to relate, enjoy inexpressible rapture. For then the spirit sees what it never has recognized; it sees the air between the heart and itself radiantly beaming." This light, the hermits declare, is the light of God that was visible to the young men on Tabor.

Yoga-Sutra (Patanyali, I, 36) says: "Or that sorrowless condition of mind, full of light (would conduce to samadhi)." And the commentator Mánilal Nabubhai Dvivedi remarks upon this:

" The light here referred to is the light of pure sattva. When the mind is deeply absorbed in that quality, then, indeed, does this condition of light which is free from all pain follow. Vachaspatimisra remarks that in the heart there is a lotus-like form having eight petals and with its face turned downward. One should raise this up by rechaka (exhalation of the breath) and then meditate upon it, locating therein the four parts of the pranava, viz., a, u, m, and the point in their several meanings. When the mind thus meditating falls in the way of the susumna, it sees a perfect calm light like that of the moor of the sun, resembling the calm ocean of milk. This is the jyotis, light, which is the sure sign of complete sattva. Some such practice is here meant. . . ." The similarity to the instruction of the Abbot Simeon is evident.

The light and sun symbolism in alchemistic writings is everywhere used; yet gold also = sun, indeed the same sign ☉ serves for both. I should like to call attention incidentally to a beautiful use of the sun symbol in " Amor Proximi," which differs slightly from the more restricted gold symbolism. On p. 32 ff. we read:  " See Christ is not outside of us, but he is intimately within us all, but locked up, and in order that he may unlock that which is locked up in us, did he once become outwardly visible, as a man such as we are, the hard sin enclosure excepted, and of this the ☉ in this world is the true copy, which quickly convinced the heathens from the beginning of the world that God must become man even as the light of nature has become a body in the ☉. Now the ☉ is not alone in the firmament outside of all other creatures, but it is much more in the center of all creatures but shut up, but the external ☉ is as a figure of Christ, in that it unlocks in us the enclosed ☉, as its image and substance, just as Christ does, through his becoming man, also unlock in us the image of God. For were this not so, then the sphere of the earth would approach in vain to the ☉ in order to derive its power from it, and nothing at all would grow from the accursed ▽. [The symbol ▽ means earth.] So the ▽ shows us that inasmuch as it approaches near to the ☉ it is unlocked, so we, too, approaching Christ, shall attain again the image of God; then at the end of time this ▽ will be translated into the point of the sun [in Solis punctum] [Cf. what has been said about the point in the ☉.]; and still farther on: " Ye see that the ▽ turns to the sun, but the reason ye know not; if the earth had not in the creation gone out of the Solis punctum, it could not have turned and yearned according to its magnetic manner, so this turning around shows us that the world was once renewed, and in its beginning, as ☉ is punctum; it desires to return, and its rest will be alone in that; therefore the soul of man is also similarly gone out of the eternally divine sun, towards which it also yearns. . . ."

Our parable, to which I should like now to revert, appears in a new light. It would be a waste of time to lead the reader once more through all the adventures of the wanderer.   He again, without difficulty, will find all the aforesaid elements in the parable, and will readily recognize the introversion and rebirth.   I therefore pick out for further consideration only a few particular motives of the parable or alchemy which seem to me to require special elucidation.

We should not forget the singular fact that after the introversion, at the beginning of the work of rebirth, a deluge occurs.   This flood takes place not merely in the alchemistic process (when the bodies undergo putrefaction in the vessel and become black), but we see the mythic deluges coming with unmistakable regularity at the same time, i.e., after the killing of the original being (separation of the primal parents, etc.), and before the new creation of the world by the son of God.   Stucken (SAM., p. 123): "We see corroborated . . . what I have already emphasized, that on the appearance of the flood catastrophe the creation of the world is not yet finished.   Even before the catastrophe there was indeed an earth and life on it, but only after the flood, begins the forming of the present Cosmos.   Thus it is in the germanic Ymir-saga, and in the Babylonian Tiamat-saga, in the Egyptian and likewise in the Iranian."   What may the flood be in the psychological sense.   Dreams and poetry tell us, in that they figure the passions in the image of a storm-tossed sea.   After the introversion, whose perils have already been mentioned, there is always an outbreak of the passions.   Not without consequences is the

Stone of the Deeps elevated, which locks the
prison of the subterranean powers. (Cf. Book of
Enoch, x, 5, and passim.)   The point is to seize the
wildly rushing spirits and to get possession of their
powers without injury.   The entire inundation must,
in the philosophical vessel, be absorbed by the bodies
that have turned black, and then it works on them
for the purpose of new creation, fructifying them
like the floods of water upon the earth.   It does no
damage to the materia only then, when it is actually
black (stage of victory).   If this happens, it (the
materia) is in contrast to the waters raging over it,
like an ocean which suffers no alteration by the influx
of waters.   " Like an ocean that continually fills itself and yet does not overflow its boundaries, even
with the inflowing waters, so the man acquires calm,
into whom all desires flow in similar wise, and not
he who wantonly indulges his desire."   [Bhag. Gita,
II, 70.  Latin: translated by Schlegel:  German
[Schroeder].

"Wer wie das Meer in das die Wasser strömen
    Das sich anfüllet und doch ruhig dasteht
    Wer so in sich die Wünsche lässt verschwinden,
    Der findet Ruhe — nicht wer ihnen nachgibt."

Above I have compared the lion of the parable to
the Sphinx of Œdipus, and on the other hand, it appears from later deliberation that it (the lion) must
be the retrogressive element in men, which is to be
sacrificed in the work of purification.   Now I find
several remarks of Jung (Psychology of the Uncon-

scious) that mediate very well between both ideas. Even if I do not care to go so far as to see in the animal only the sexual impelling powers, but prefer to regard it rather as the titanic part of our impulses, I find the conception of the author very fortunate. The Sphinx, that double being, symbolizes the double natured man, to whom his bestiality still clings. Indeed it is to be taken exactly as a functional representation of the development of reason out of the impulses (human head and shoulders growing out of an animal body).

The homunculus motive would likewise have to be regarded in a new light. I have said that the mystic was his own father; he creates a new man (himself) out of himself with a merely symbolic mother, therefore with peculiar self-mastery, without the coöperation of any parents. That means the same thing as the artificial creation of a man. We recognize therefore the anagogic significance of the homunculus, the idea of which we found closely interwoven with alchemy in general. This connection also has not escaped Jung, though he takes it onesidedly and draws a too far-reaching conclusion. He points to the vision of Zosimos, where, in the hollow of the altar he finds boiling water and men in it, and remarks that this vision reveals the original sense of alchemy, an original impregnation magic, i.e., a way in which children could be made without a mother. I must observe that the hermetic attempt to get back to Adam's condition has some of the homunculus phantasy in it. Adam was regarded

as androgyne, a being at once man and woman, but
sufficient in himself alone for impregnation and procreation. Welling says in his Opus mago-cabalisti-
cum, " This man Adam was created, as the scripture says, i.e., of the male and female sex, not two
different bodies but one in its essence and two in its
potentiality, for he was the earth Adamah, the red
and white ♁, the spiritual ☉ and ☽, the male and
female seed, the dust of the Adamah from Schama-
jim, and therefore had the power to multiply himself
magically (just as he was celestial) which could not
indeed have been otherwise, unless the essential mas-
culinity and femininity were dissociated." I am reminded in this connection that Mercury is also bisexual; the " materia " must be brought into the
androgynic state " rebis." The idea of hermaphroditism plays a well known, important part in mythology also.

*     *     *

We have explained why phantasy creations carry
two meanings, the psychoanalytic and the anagogic,
apparently fundamentally different, even contradictory, and yet, on account of their completeness, unde-
niable. We have found that the two meanings correspond to two aspects or two evolutionary phases
of a psychic inventory of powers, which are attached
as a unity to symbolic types, because an intro-determination can take place in connection with the sublimation of the impulses. When we formulated the
problem of the multiple interpretation, we were
struck with the fact that besides the two meanings

that were nominally antipodal in ethical relations, there was a third ethically indifferent, namely, the natural scientific. Apart from the fact that I have not yet exhausted the anagogic contents of our material and so must add a number of things in the following sections, I am confronted with the task of elucidating the position of the nature myth portion. That will necessarily be done briefly.

In the case of alchemy the natural scientific content is chemistry (in some degree connected with physics and cosmology), a fact hardly requiring proof. The alchemistic chemistry was not, to be sure, scientific in the strict modern sense. In comparison with our modern attitudes it had so much mythical blood in it that I could call it a mythologically apperceiving science, wherein I go a little beyond the very clearly developed conception of Wilhelm Wundt (Volkerps. Myth. u. Rel.) regarding mythological apperception, from a desire for a more rigid formulation, but without losing the peculiar concept of the mythical or giving it the extension it has acquired with G. F. Lipps. Alchemy's myth-like point of view and manner of thinking is paralleled by the fact that it was dominated by symbolic representation and the peculiarities that go with it. [The concept of the symbol is here to be taken, of course, in the wider sense, as in my papers on Symbolbildung (Jb. ps. F., II–IV).]

The choice of a symbol is strongly influenced by what strongly impresses the mind, what moves the soul, whether joyful or painful, what is of vital in-

terest, in short, whatever touches us nearly, whether
consciously or unconsciously. This influence is
shown even in the commonplace instances, where the
professional or the amateur is betrayed by the manner of apperceiving one and the same object. Thus
the landscape painter sees in a lake a fine subject, the
angler an opportunity to fish, the business man a
chance to establish a sanitarium or a steamboat line,
the yachtsman a place for his pleasure trips, the heat
tormented person a chance for a bath, and the
suicide, death. In the symbolic conception of an
object, moreover (which is much more dependent
on the unconscious or uncontrolled stimulation of the
phantasy that shapes the symbol), the choice from
among the many possibilities can surely not fall upon
such images as are unsympathetic or uninteresting
to the mind. Even if we consciously make comparisons we think of an example mostly from a favorite
and familiar sphere; when something " occurs " to
us there is already evidenced some part of an unconscious complex. This will become elaborated in the
degree that the phantasy is given free play.

The raw product then, of the symbol-choosing
phantasy of the individual (" raw," i.e., not covered
for publicity with a premeditated varnish) bears
traces of the things that closely concern the person
in question. (" Out of the fulness of the heart the
mouth speaketh "— even without premeditation.)
If we now start from a spiritual product which is
expressed in symbols (mythologically apperceived),
and whose author we must take to be not an individ-

ual man but many generations or simply mankind, then this product will, in the peculiarities of the selection of the symbol, conceivably signify not individual propensities but rather those things that affect identically the generality of mankind. In alchemy, which as a mythologically apperceiving science is completely penetrated by symbols, we regard as remarkable in the selection of symbols, the juxtaposition of such images as reflect what we have, through psychoanalysis, become acquainted with, as the " titanic " impulses (Œdipus complex). No wonder! These very impulses are the ones that we know from psychoanalytic investigations as those which stand above all individual idiosyncracies. And if we had not known it, the very circumstances of alchemy would have taught us.

The familiar scheme of impulses with its " titanic " substratum, which is necessarily existent in all men (although it may have been in any particular case extraordinarily sublimated) comes clearly to view in individual creations of fancy. It must be found quite typically developed, however, where a multitude of men (fable making mankind) were interested in the founding, forming, polishing and elaborating of the symbolic structure. Such creations have transcended the merely personal. An example of this kind is the " mythological " science of alchemy. That we are repelled by the retrograde perspective of the types residing in its symbols (and which often appear quite nakedly) comes from the fact that in the critic these primal impulse forms have experienced a strong repression, and that their re-emergence meets a strong resistance (morality, taste, etc.).

The much discussed elementary types have therefore insinuated themselves into the body of the alchemistic hieroglyphics, as mankind, confronted with the riddles of physico-chemical facts, struggled to express a mastery of them by means of thought. The typical inventory of powers, as an apperception mass, so to speak, helped to determine the selection of symbols. A procedure of determination has taken place here similar to that we might have noticed in the coincidence of material and functional symbolism in dreams. Here again appears the heuristic value which the introduction of the concept of the functional categories had for our problem.

The possibility of deriving the " titanic " and the " anagogic " from the alchemistic (often by their authors merely chemically intended) allegories is now easily explained. We can work it out, because it was already put in there, even if neither in the extreme form of the " titanic " (i.e., the retrograde aspect), nor in that of the " anagogic " (the progressive aspect), but in an indeterminate middle stage of the intro-determination. What gave opportunity for this play of symbolism was an effort of intelligence directed toward chemistry. The chemical content in alchemy is, so to speak, what has been purposely striven for, while the rest came by accident, yet none the less inevitably. So then natural philosophy appears to be the carrier, or the stalk on which the titanic and the anagogic symbolism blossoms. Thus it becomes intelligible how the alchemistic hieroglyphic aiming chiefly at chemistry, adapted itself through and through to the hermetic anagogic educational goal, so that at times and by whole groups, alchemy was used merely as a mystical guide without any reference to chemistry.

What we have found in alchemy we shall now apply to mythology where analogous relations have been indicated. [The apperception theory here used should not be confused with the intellectual theory (of Steinthal) which Wundt (V. Ps., IV², pp. 50 ff.) criticised as the illusion theory. I should be more inclined to follow closely the Wundtian conception of the " mythological apperception " (ibid., pp. 64 ff.) with particular emphasis on the affective elements that are to work there. With Wundt, the affects are really the " actual impulse mainsprings " and the most powerful stimuli of the phantasy (ib., p. 60). " The affects of fear and hope, wish and desire, love and hate, are the widely disseminated sources of the myth. They are, of course, continually linked with images. But they are the ones that first breathe life into these images." I differ from Wundt in that I have more definite ideas of the origin of these affects, by which they are brought into close connection with the frequently mentioned elementary motives.] Modern investigation of myths has, in my opinion, sufficiently shown that we are here concerned with a nucleus of natural philosophy (comprehension of astral and even of meteorological processes, etc.) around which legendary and historical material can grow. As has been shown by two fairy tales and as I could have abundantly shown from countless others, the psychoanalytic and the anagogic interpretations are possible alongside of the scientific. [We can criticise Hitchcock for having in his explanations of fairy tales considered them only in their most developed form, and not bothered about their origin and archaic forms. And as a matter of fact the more developed forms permit a very much richer anagogic interpretation than the archaic. But that is no proof against the interpretation, but only establishes their orientation in the development of the human spirit. The anagogic interpretation is indeed a prospective explanation in the sense of an ethical advance. Now the evolution even of fairy tales shows quite clearly a progression towards the ethical; and inasmuch as the ethical content of the tale grows by virtue of this evolution, the anagogic explanation is in the nature of things able to place itself in higher developed tales in correspondingly closer connection with mythical material.] I adduce here only one example, namely the schema that Frobenius has derived from the comparison of numerous sun myths. The hero is swallowed by a water monster in the west [the sun sets in the sea]. The animal journeys with him to the east (night path of the sun apparently under the sea). He lights a fire in the belly of the animal and cuts off a piece of the pendant heart when he feels hungry. Soon after he notices that the fish is running aground. (The reillumined sun comes up to the horizon from below.) He begins immediately to cut his way out of the animal, and then slips out (sunrise). In the belly of the fish it has become so hot that all his hair has fallen out. (Hair probably signifies rays.) Quite as clear as the nature myth purport, is the fact that we have a representation of renegeration, which is quite as conceivable in psychoanalytic as in anagogic explanation.

Now I cannot approve of the attempt of many psychoanalysts to treat as a negligible quantity or to ignore altogether the scientific content (nature nucleus) of the myths which has been so well substantiated by the newer research, even though it is not so well established in the details. [I have uttered a similar warning in Jb. ps. F., IV (Princip. Anreg.) and previously, in Jb. ps. F., II (Phant. u. Myth), have advocated the equality of the natural philosophical and the psychological content. Now I observe with pleasure that very recently an author of the psychoanalytic school is engaged on the very subject that I have recommended as so desirable. Dr. Emil F. Lorenz, in the February number of *Imago*, 1913, treats the " Titan Motiv in der allgemeinen Mythologie " in a manner that approaches my conception of it. In the consideration of human primal motives as apperception mass, there is particularly revealed a common thought in the primitive interpretation of natural phenomenon. Unfortunately the article appeared after this book was finished. So even if I am not in a position to enter

into this question, I will none the less refer to it
and at the same time express the hope that Lorenz
will further elaborate the interesting preliminary
contribution, communicated in the form of apho-
risms, as he terms it.]   The inadmissibility of these
omissions arises from the vital importance and grip-
ping effect of the objects thus (i.e., mythologically)
regarded by humanity (e.g., of the course of the sun,
so infinitely important for them in their dependence
upon the moods of nature).   If then, on the one
hand, it will not be possible for the psychoanalyst to
force the nature mythologist out of his position and
somehow to prove that any symbol means not the sun
but the father, so on the other hand the nature
mythologist who may understand his own interpretations so admirably, must not attack the specifically
psychological question: why in the apperception of
an object, this and not that symbolic image offers itself to consciousness.   So, for instance, why the sun-
set and sunrise is so readily conceived as a swallowing and eructation, or as a process of regeneration.
Yet Frobenius (Zeitalt. d. Sonneng., I, p. 30) finds
the symbolism " negligible."

It is also conceivable that the obtrusive occurrence of incest, castration of the father, etc., should
make the mythologists ponder.   It was bias on the
part of many of them to be unwilling to see the psychological value of these things.   I must therefore
acknowledge the justice of Rank's view when he
(Inz-Mot., p. 278) says in reference to the Œdipus
myth (rightly, in all probability, interpreted by

Goldziher as a sun myth): "Yet it is indubitable
that these ideas of incest with the mother and the
murder of the father are derived from human life,
and that the myth in this human disguise could never
be brought down from heaven without a corresponding psychic idea, which may really have been an unconscious one even at the time of the formation of
the myth, just as it is with the mythologists of to-
day."

And in another passage (pp. 318 ff.): "While
these investigators (astral and moon mythologists)
would consider incest and castration operative in an
equal or even greater degree than we do, as the
chief motives in the formation of myths in the celestial examples only, we are forced by psychoana-
lytical considerations to find in them universal primitive human purposes which later, as a result of the
need of psychological justification, have been projected into the heavens from which our myth interpreters wish in turn to derive them.   [Whether such
a need of justification has had a share in the formation of myths appears to me doubtful or at any rate
not demonstrable.   At all events in so strongly em-
phasizing these unnecessary assumptions and conceiv-
ing the projection upon heaven of the mundane psychological primal motives as an act of release, we
hide the more important cause for concerning ourselves with heaven, namely the already mentioned
vital importance of the things that are accomplished
there.   Now the fact that the primal motives co-
operate in the symbolical realization of these things,

implies no defense directed against them. A better
defense would be to repress them in symbolism than,
as really happens, to utilize them in it.] These interpreters, for example, have believed that they
recognized in the motive of dismemberment (castration) a symbolic suggestion of the gradual
waning of the moon, while the reverse is for us un-
doubted, namely, that the offensive castration has
found a later symbolization in the moon phases.
Yet it argues either against all logic and psychology,
or for our conception of the sexualization of the universe, that man should have symbolized so harmless
a phenomenon as the changes of the moon, by so
offensive a one as the dismemberment or castration
of the nearest relative. So the nature mythologists
also, and Siecke in particular, have thought that
primitive man has "immediately regarded" the (to
him) incomprehensible waning of the moon as a dismemberment, while this is psychologically quite un-
thinkable unless this image, which is taken from
earthly life, should have likewise originated in
human life and thought (phantasy).

It is indeed never conceivable that men would have
chosen for the natural phenomenon exactly these
titanic symbols, if these had not had for them a special psychic value, and therefore touched them
closely. If any one should object that they would
not have "chosen" them (because they did not purposely invent allegories, as was formerly thought),
I should raise the contrary question: Who has chosen
them? I will stick to the word "choose" for a

choice has taken place. But the powers that arranged this choice lived and still live in the soul of man.

The conception advocated by me gives their due to the nature mythologists just as much as to the psychologists that oppose them. It reinstates, moreover, a third apparently out-worn tendency [the so-called degeneration theory] that sees in the myth the veiling of ancient priestly wisdom. This obsolete view had the distinction that it placed some value, which the modern interpreters did not, on the anagogic content of the myths (even if in a wrong perspective). The necessity of reckoning with an anagogic content of myths results from the fact that religions with their ethical valuations, have developed from mythical beginnings. And account must be taken of these relations. In the way in which the older interpretations of myths regarded the connection, they pursued a phantom, but their point of view becomes serviceable as soon as it reverses the order of evolution. It is not true that the religious content in myths was the priestly wisdom of antiquity, but rather that it became such at the end of the development. My conception shows further that the utmost significance for the recognition and comparison of the motives (corresponding to the psychological types) attaches to the material so brilliantly reconstructed by Stucken and other modern investigators, but not the convincing evidence which some think they find there for the migration theory, as against the theory of elementary thoughts.

With regard to the possibly repellent impression derived from the notion of an unconscious thought activity of the myth forming phantasy, I should like to close with these words of Karl Otfried Müller: " It is possible that the concept of unconsciousness in the formation of myths will appear obscure to many, even mysterious . . . but is history not to acknowledge the strange also, when unprejudiced investigation leads to it? "

## SECTION II

## THE GOAL OF THE WORK

In the preceding section the symbolism and the psychology of the progress of the mystic work has been developed more or less, but certainly not to the end. Regeneration is evidently the beginning of a new development, the nature of which we have not yet closely examined. Nothing has yet been said definitely about the later phases of the work and about its goal. I am afraid that this section, although it is devoted chiefly to the goal of the work, cannot elucidate it with anything like the clearness that would be desirable. To be sure the final outcome of the work can be summed up in the three words: Union with God. Yet we cannot possibly rest satisfied with a statement that is for our psychological needs so vague; we must endeavor to comprehend the intimate nature of the spiritual experiences that we have on the journey into the unsearchable; although I must at the outset point out that at every step by which the symbolism of the mystics leads us towards regeneration, we run the risk of wandering away from psychology, and that in the following we shall all too soon experience these deviations. We shall have to transplant ourselves uncritically at times, into the perceptual world of the hermetics, which is, of course, a mere fiction, for in order to do it rightly we should have to have a mystical development behind us [whatever this may be]; one would have to be himself a " twice born." One thing can be accepted as true, that a series of symbols that occur with striking agreement among all mystics of all times and nations is related to a variety of experiences which evidently are common to all mystics in different degrees of their development, but are foreign to the non-mystics (or more exactly to all men, even mystics, who have not attained the given level).

With this premise I will take up the question of the goal of alchemy (mysticism).   In this I follow in general the train of thought of Hitchcock, without adhering closely to his exposition.   (I cite H. A. = Hitchcock, Remarks upon Alchemy.)

The alchemistic process is, as the hermetics themselves say, a cyclical work, and the end resides to a certain degree in the beginning.   Here lies one of the greatest mysteries of the whole of alchemy, although the meaning of the language is to be understood more or less as follows.   If, for example, it is said that whoever wishes to make gold must have gold, we must suppose that the seeker of truth must *be* true (H. A., p. 67); that whoever desires to live in harmony with the conscience must be in harmony with it, and that whoever will go the way to God, must already have God in himself.   Now when the conscience, wherein the sense of right and justice has existence, becomes active under the idea of God, it is endowed with supernatural force and is then, as I understand it, the alchemists' philosophical mercury and his valued salt of mercury. It is no less his sovereign treacle, etc. (H. A., p. 53). The progress of the work points to some kind of unity as the goal which, however, very few men attain except in words (H. A., p. 157). The hermetic writers set up the claim to a complete agreement in their teachings, but this agreement is restricted to some principles of vital significance in their doctrine, which have reference almost exclusively to a definite practice; probably to a complete setting to work of the consciousness of duty, which is what Kant claims to do with his categorical imperative: " An unreasoning, though not unreasonable, obedience to an experienced, imperious sense of duty, leaving the result to God; and this I am disposed to call the Way."

> Do thy duty!
> Ask not after the result of thy doing!
> Without dependence thereon carry out that which is thy
>     duty!
> Whoever acts without attachment to the world, that man
>     attains the loftiest goal.

And the like in many places in the Bhagavad-Gita.]

Now the end is perhaps the fruit of this obedience. It may be that the steady preservation of the inward unity, which regards with composure all external vicissitudes, leads man finally to some special experience, by which a seal of confirmation is set upon what was first a mere trust in the ultimate blessing of rectitude (H. A., p. 128). The hermetic philosophers would have the conscience known as the Way or as the base of the work, but with regard to the peculiar wonder work of alchemy (transmutation) they place the chief value on love; it effects the transformation of the *subject* into the *object* loved (H. A., p. 132).

Arabi: " It is a fundamental principle of love that thou becomest the real essence of the beloved (God) in that thou givest up thy individuality and disappearest in him. Blessedness is the abiding place of the divine and holy joy." (Horten, Myst., I, p. 9.)

Similarly we find in the yoga primers that the spirit, by sinking into an object of perception, becomes identical with the object. The object need not be the very highest, but a gradation is possible. Arabi, too, recognizes a gradation of objects, as they correspond, as correlates of sinking or surrender, to the different mystical states. [Colors, etc., of alchemy.] Two passages of Arabi may be quoted: " My heart is eligible for every form [of the religious cult]; for it is said that the heart (root: kalaba = overturn, to alter oneself) is so called from its continual changing. It changes in accordance with the various (divine) influences that it feels, according to the various states of the mystical illumination. This variation of experiences is a result of the variation of the divine appearances, which occur in its inmost spirit. The law of religion (theology) speaks of this phenomenon as the changing and metamorphizing in the forms (of living and being). Gazelles are the objects of the mystic's love. In one of his poems he says: "And surrender yourselves to play in the manner of lovely maidens with buxom breasts and enjoy the luxuriant willows in the manner of the female gazelles." In his commentary on this passage he says: "'Play' denotes the various states of the mystic, to which he is advanced when he passes from one divine name to another." (Horten, Myst., I, pp. 11, 13, ff.)

It is the ethical ideal of the mystic, more and more to put off the limited ego, and to take on in its place the qualities of God, in order to become God.

When with Arabi the theme of an ode is "Through asceticism, fervent yearning after God and patience in suffering, man becomes God or acquires divine nature" (Horten, Myst., I, p. 16), then this goal is identical with that of the alchemistic transmutation; the base metal acquires (after purification, refining, etc.) by virtue of the tincturing with the Philosopher's Stone the nature of gold, i.e., the divine nature.

But patient effort is requisite. Precipitancy is as great an evil as inactivity. It is, to use the language of the alchemists, just as bad to scorch the tender blossoms by a forced and hasty fire (that in spite of its intensity may be merely a straw fire), as to let go out the fire which should be continuously kept alight, and to let grow cold the materia. The process of distillation is to be accomplished slowly, so that the spirits may not escape.    That which rises as steam through the " heating " in the " receptacle " (i.e., in man) is the soul rising into the higher regions.    Distilling like rain drops [destillare = drop down], it brings each time to the thirsting materia a divine gain.    But this process is not to be overdone, for the thirsting earth must be gently instilled with the heavenly moisture of the water of life: the process of " imbibition."

The metallic subject must be gently dissolved in its own natural water (conscience), not with powerful media, not with corroding acids, which the foolish employ in order to reach the goal in a hurry, for by such means he either spoils the materia or produces a merely superficial action.    Senseless asceticism and the like are just as objectionable as the impetuous enthusiasm (which we called straw fire here).    The ethical work of alchemy as of common life is a sublimation; it is important that the materia takes up at any time only as much as it can sublimate.    We may also conceive it in this way. The materia is to be moistened only with the water that it can utilize after the solution has taken place (i.e., keep in enduring form, absorb into their nature).    Compare in this connection the words of Count Bernhard von Trevis:    " I tell you assuredly that no water dissolves any metallic spices by a natural solution, save that which abides with them in matter and form, and which the metals themselves, being dissolved, can recongeal."    (H. A., pp. 189 ff.)

The passage "slowly and quite judiciously" of the Smaragdine tablet will now be fully appreciated.

The desired completion or oneness should be a state of the soul, a condition of being, not of knowing. The means that lead to it presuppose in the neophyte something analogous to religious faith, and because the conditions of the mastery appear to the neophyte to contradict nature or each other, the mystical experiences that are derived from it are called "supernatural." The "supernatural" is, however, only an appearance, which results when we conceive nature too narrowly, as when we see in her merely the totality of bodies. If we mean by nature the possibility of life and activity, then that which appears supernatural must be counted as nature. The expressions natural and supernatural are but means of the thinking judgment, they are preliminaries which have a certain justification but only so long as they are an expression for a stage of knowledge. The initially supernatural resolves itself in nature, or better, Nature is raised to divinity. If the natural and the supernatural are symbolized, the one being described as sulphur and the other as mercury, then the disciples of philosophy, under the obligation to think things and not merely names, are finally brought, during the process of search, to a recognition of the inseparableness of both in a third something which may be called sun; but as all three are recognized as inseparably one, the termini can change places until finally an inner illumination takes place. "Those that have never had this experience are apt to decry it as imaginary, but those who enter into it know that they have entered into a higher life, or feel themselves enabled to look upon things from a higher point of view.    To use what may seem to be a misapplication of language: it is a supernatural birth, naturally entered upon." (H. A., p. 229.)    When the alchemists speak of philosophical mercury and philosophical gold, they mean something in man and something in God that finally turns out to be the One.    " By this symbolism the alchemists escape the difficulty of treating the subject in ordinary language.    The learner must always return to nature and her possibilities for the sense of the derived symbols, and to it the hermetic masters also continually direct him." (H. A., pp. 232 ff.)    If the true light has risen in the hearts of the seekers, kindled from within (although apparently by a miracle from without) " the sulphur and mercury become one, or are seen to be the same, differing only in a certain relation; somewhat as the known and the unknown (and the conscious and the unconscious) are but one, the unknown decreasing as the known increases, and vice versa." (H. A., p. 235.)

One alchemist teaches: " Consider well what it is you desire to produce, and according to that regulate your intention.    Take the last thing in your intention as the first thing in your principles. . . . Attempt nothing out of its own nature [then follow parables that grapes are not gathered from thistles, etc.].    If you know how to apply this doctrine in your operation as you ought, you will find great benefit, and a door will hereby be opened to the discovery of greater mysteries." Actually there is a greater difference between one who seeks what he seeks as an end, and one who seeks it as a means to an end. To seek knowledge for riches is a very different thing from seeking riches (or independence) as an instrument of knowledge. In the study in question the means and the end must coincide, i.e., the truth must be sought for itself only. (H. A., p. 238.) In the book, " De Manna Benedicto," we read: " Whoever thou art that readest this tractate, let me exhort thee that thou directest thy understanding and soul more toward God for the keeping of his commandments, than toward love of this art [sc. its external portions], for although it be the only, indeed the whole wisdom of the world, it is yet powerless in comparison with the divine wisdom of the soul, which is the love towards God, in the keeping of his commandments. . . . Hast thou been covetous, profane one? Be thou meek and pious and serve in all lowliness the glorious creator; if thou art not determined to do that, thou art employed in trying to wash an Ethiop white."

Desire is, as some ancient philosophers think, the root of all affects, which manifest themselves in pairs. Joy corresponds to desire fulfilled, sorrow to the obstructed or imperiled fulfillment; hope is the expectation of fulfillment, fear the opposite, etc. All the pairs of opposites are in some degree superficial, something that comes and goes with time, while the essential remains, itself invisible and without re-

lation to time — a perpetual activity, an ever enduring conation as it was formerly called.  (It is the
libido of the psychoanalysis.   In its manifestations
it is subjected to bipolarity, as Stekel has named the
inevitable pairs of opposites.)

The pairs of opposites have been noticed in the
Hindu doctrine of salvation exactly as in alchemy.
Alchemistic hieroglyphics we know are rich in [am-
biguous] expressions for a hostile Dyas (couple),
with whose removal a better condition first com-
mences, although at the outset it is actually requisite
for the achievement of the work.   In the Bhagavad-
Gita the pairs of opposites play a great part.   The
world is full of agony on account of the pairs of
opposites, which are to be found everywhere.   Heat,
cold; high, low; good, evil; joy, sorrow; poor, rich;
young, old; etc.   The basis of the opposites is
formed by the primal opposition Rajas-Tamas.   To
escape from it in recognizing the true ego as superior to it and not participating in it, is the foremost
purpose of the effort toward salvation.   So whoever has raised himself above the qualities of substances is described as having escaped from opposites.

" Contact of atoms is only cold and warm, brings pleasure
      and pain,
They come and go without permanency — tolerate them O
      Bharata.
The wise man, whom these do not affect, O mighty hero,
Who bears pain and pleasure with equanimity he is ripening
      for immortality."   (II, 14 ff.)

The spirit, the true ego, is raised above the agitation of the qualities of nature:

"Swords cut him not, fire burns him not,
Water wets him not nor does the wind wither him.
Not to be cut, not to burn, not to get wet, not to be withered,
He is constant, above everything, continuous, eternal immovable." [II 23 ff.]

This characterization sounds almost like the description of the mercury of the philosophers, which is indestructible, a water that does not wet, a fire that does not consume.

Hermes on the human soul: "The accidents residing in the material substances have never sympathized with each other, but on the contrary have always been in opposition and in mutual conflict. Guard thyself O soul from them and turn away from them. . . . Thou O soul art of one nature, but they are manifold; thou art but one with thyself; they are, however, in conflict with each other. [Psychoanalytically regarded, to the soul is here assigned the property which is desired but is not present, while that which is undesired but actually present in the soul (inclination and disinclination) is projected into the external world.] . . . How long O soul wilt thou yet be needy, and flee from every sensation to its opposite, now from warmth to cold, now from cold to warmth, now from hunger to satiety, now from satiety to hunger?" (Fleischer Herm. a. d. Seele, pp. 14 ff.)    "Be thou O soul regardful of the behavior in this world, yet not as a child without understanding who when one gives him to eat and acts leniently towards him is satisfied and cheerful, but when one treats him severely cries and is bad, indeed begins to weep while laughing and when he is satisfied begins again to be bad.   This is not worthy of approbation but rather a mongrel and blameworthy behavior.   The world O soul, is so organized as to unify exactly these opposites; good and evil, weal and woe, distress and comfort, and contains types of ideas that have the effect of waking the soul and making it aware of itself, so that as a result it gains reason that illumines and consummates knowledge, i.e., wisdom and knowledge of the true nature of things.   For this purpose alone has the soul come into the world, to learn and experience; but it is like a man that comes to a place to become acquainted with it and know its conditions, but then gives up the learning, inquiring and collecting of information, and diverts his spirit by reaching after luxury and the enjoyment of other things, and in so doing forgets to acquire that which he was to strive for."   (L. c., pp. 8 ff.)

I return to the psychological point of view of our friend Hitchcock: " Desire and love are almost synonymous terms, for we love and seek what we desire, and so also we desire and seek what we love; yet neither love nor desire is by any necessary connection directed to one thing rather than another, but either under conditions suitable to it may be directed to anything.   From which it follows that it is possible to make God as the Eternal, its object, or call it truth and we may see that its enjoyment must partake of its own nature.    Now we read that it is not common for man to love and pursue the good and the true because it is the good and true; but we call that good which we desire and there lies the great mistake of life.    From all which we may see that vast consequences follow from the choice of an object of desire, which as we have said, may as easily be an eternal as a transient one.    We should be on guard against a too mechanical conception of these things.    By so doing we should depart too greatly from the point of view of the true alchemists.    One author tells of the significant advance that he made from the time when he discovered that nature works 'magically.' "    (H. A., Hitchcock's Remarks upon Alchemy, pp. 294 ff.)

Aversion and hate, the opposites of desire and love, are not independent affections but depend upon the latter.    There is only the one impulsion of demand that strives for what satisfies it and repulses what conflicts with it.    "If then desire is turned to one only eternal thing, then, since the nature of man takes its character from his leading or chief desire, the whole man is gradually converted to, or, as some think, transmuted into that one thing."    (H. A., pp. 295 ff.)

The doctrine naturally presupposes the possibility, already mentioned, of a schooling of the will, yet it will still be necessary to fix it upon a definite object. The love of the transitory finds itself deceived because the objects vanish, while the desire itself, the conation (or in psychoanalytic language the libido), continues forever.   For this everlasting desire only an everlasting object is suitable.   An object of that kind is not to be found in the external world.   We can only withdraw the outer object and offer ideals in exchange.   The moment that this withdrawal of external objects takes place the libido begins, as it were, to eject itself as an object; in the ideal we give it a nucleus for this process, in order that it may form the new object around it and water it with its own life.   So in a " magic " way a new world is formed whose laws are those of the ideal.   The formation of the new world (new earth and new heaven, new Jerusalem, etc.) occurs frequently in the symbolic language of mysticism.

The laws of the ideal and consequently of the new world are determined by the nature of the ideal. Not every one is proved everlastingly suitable.

" Those that dedicate themselves to the gods and
        fathers, pass over to the gods and fathers,
Spirit worshipers to the spirits, whoever honors me,
        comes to me."

says the Highest Being to Arjuna in the Bhagavad-Gita (IX, 25).   The mystic is in the position from the moment of regeneration, to create in himself a new world with laws that he may, to a certain extent, himself select.   Fortunate is he who makes a good selection.   Every one is the architect of his own fortune.   This is most true when after introversion the power of self determining one's own destiny is directed toward the most intensive living. The formation and cultivation of the new earth is a beginning that is rich with significant consequences. The alchemists speak of a maidenly earth or a flaky white earth (i.e., crystalline) as a certain stage in the work. This is probably the stage that we are examining now, the stage of the new, still undeveloped earth that is now to be organized (according to the conceived ideal). The soil is crystalline because the old earth was dissolved and has been freshly formed from the solution. The crystallization corresponds to regeneration. The "white earth" probably corresponds to the "white stone," which is the first stage of completion after the blacks (first mystical death, putrefaction, trituration, or contrition). In the white earth a seed is sown. We shall hear of it later.

If the work is not to make men unserviceable and is not again to bring them into conflict with the demands of life, so that all the effort would have been fruitless, the new world must be organized in such a way that it is compatible with the demands of real life. In other words, the ideal that regulates the new world must be an ethical one. The mystic who wishes to be freed from contradictions will have to follow his conscience as a guide, and not the unexplored but the explored conscience. He cannot escape it in the long run (the magicians that defy it are, as the legend informs us, finally torn to pieces by the devil); it is better for him to get upon its

side and so turn the conflict in his favor.   It appears
that this manly attitude would have a marvelous
inner concord as a result and outwardly, a remarkable firmness of character.   It is not my object to
decide what metaphysical significance the strengthening through mysticism of the ideal (God in me)
may have.

" Take,  O soul, not the unworthy and common as
a model, for such use and word will adhere to thee
finally as a nature opposed to thine own.   By this
means, however, the strong impulse itself towards
union with thy nature and to the return into thy
home goes astray.   Know that the exalted and ma-
jestic Originator of things, is himself the noblest of
all things.   Take then the noble things as a model,
in order by that means to get nearer thy Creator on
the path of elective affinity.   And know that the
noble attaches itself to the noble and the vulgar to
the common."   (Fleischer, Herm. a. d. Seele, p.
18.)

What is to be sown in the new earth is generally
called love.   A crop of love is to arise; with love
will the new world be saturated; its laws will be the
laws of love.   By love a transmutation of the subject is to take place.   One alchemist (quoted in
H. A., pp. 133 ff.) writes as follows:

" I find the nature of Divine Love to be a perfect
unity and simplicity.   There is nothing more one,
undivided, simple, pure, unmixed and uncompounded
than Love. . . .

" In the second place I find Love to be the most

perfect and absolute liberty. Nothing can move Love, but Love; nothing touch Love, but Love; nor nothing constrain Love, but Love. It is free from all things; itself only gives laws to itself, and those laws are the laws of Liberty; for nothing acts more freely than Love, because it always acts from itself, and is moved by itself, by which prerogatives Love shows itself to be allied to the Divine Nature, yea, to be God himself.

" Thirdly, Love is all strength and power. Make a diligent search through Heaven and Earth, and you will find nothing so powerful as Love. What is stronger than Hell and Death? Yet Love is the triumphant conqueror of both. What more formidable than the wrath of God? Yet Love overcomes it, and dissolves and changes it into itself. In a word, nothing can withstand the prevailing strength of Love: it is the strength of Mount Zion, which can never be moved.

" In the fourth place: Love is of a transmuting and transforming nature. The great effect of Love is to turn all things into its own nature, which is all goodness, sweetness, and perfection. This is that Divine power which turns water into wine; sorrow and anguish into exulting and triumphant joy; and curses into blessings. Where it meets with a barren and heathy desert, it transmutes it into a paradise of delights; yea, it changeth evil into good, and all imperfection into perfection. It restores that which is fallen and degenerated to its primary beauty, excellence and perfection. It is the Divine Stone, the White Stone with the name written upon it, which no one knows but he that hath it.    [Cf. Rev. II, 17. " He that hath an ear, let him hear what the Spirit saith unto the churches; To him that overcometh will I give to eat [nutritio] of the hidden manna, and will give him a white stone, and in the stone a new name written, which no man knoweth saving he that receiveth it."    Also III, 12:    " Him that overcometh will I make a pillar in the temple of my God, and he shall go no more out: and I will write upon him the name of my God, and the name of the city of my God, which is new Jerusalem, which cometh down out of heaven from my God: and I will write upon him my new name."    Cf. also XIX, 12, and XXI, 2.    The White Stone with the new name is also joined with the new earth.    Because of this it is important that the new Jerusalem is " prepared as a bride adorned for her husband."]    In a word, it is the Divine Nature, it is God himself, whose essential property it is to assimilate all things with himself; or [if you will have it in the scripture phrase] to reconcile all things to himself, whether they be in Heaven or in Earth; and all by means of this Divine Elixir, whose transforming power and efficacy nothing can withstand. . . ."    (H. A., pp. 133 ff.)

At the end of the work there ensues the union of sun and moon, typifying God and man.    As in the Vedanta the teaching of the holy books of India, the Upanishads, so in alchemy, the difference between the one soul and the All Soul is of no importance.

For every one who succeeds in overcoming the fundamental error, in which we are all implicated, the difference vanishes, and the two things previously separated coalesce.  In reality there is only the one thing: God.

Irenæus writes: ". . . The fire of nature assimilates all that it nourishes to its own likeness, and then our mercury or menstruum vanishes, that is, it is swallowed by the solar nature [The soul of man dissolves and is taken up by the divine or All Soul] and all together make but one universal mercury [All Soul] by intimate union.  And this mercury is the material principle of the Stone; for formerly, when it was compounded of three mercuries, [namely, when they thought they had to distinguish spirit, soul and body, or some other division in it] then Soul, world and God were, for example, to be thought of, or as they are called in Soeta-svatara-Upanishad V, Enjoyer, Object of Enjoyment, and Inciter.

> As eternal cause contains that trinity.
> Whoever finds in it the Brahma as the kernel,
> Resolves himself in it as a goal, and is freed from birth."

Cf. also Deussen, Syst. d. Ved., p. 232, and Sutr. d. Ved., pp. 541 ff.: " Frequently we are told of the connection of the highest with the individual soul, and then again of a splitting up [conditioned by them] inside the Brahma, by virtue of which their two parts are mutually opposed and limited.  Both of these things happen, however, only from the

standpoint of the distinctions [upadhi]. . . .
There were two which were superficial (in that they
formed an unjustified opposition) and the third essential to Sol and Luna only, not to the Stone; for
nature would produce these two out of it by artificial decoction. . . . [These distinctions depend on
ignorance, after throwing off which the individual is
one with the highest.   The connection of the individual soul with Brahman is in truth its entering
into its own self, and the division in Brahma is as
unreal as that between space in general and space
within the body.]   But when the two perfect bodies
are dissolved [prepared for the mystical work] they
are transmuted with the mercury that dissolved them,
and then there is no more repugnancy in it; then
there is no longer a distinction between superficial
and essential.   And this is that one matter of the
stone, that one thing which is the subject of all wonders.   When thou art come to this then shalt thou
no more discern a distinction between the Dissolver
[God] and the dissolved [soul] . . . and the color
of the ripe sulphur [the divine nature] inseparably
united to it will tinge your water [soul]."   Irenæus
says that the two bodies, Sol and Luna, are compared by the alchemists to two mountains, first because they are found in mountains, and second by
way of opposition: " For where mountains are
highest above ground, there they lie deepest under-
ground," and he adds: " The name is not of so
much consequence, take the body which is gold [i.e.,
here the consummate man] and throw it into mer-

cury, such a mercury as is bottomless [infinite], that is, whose center it can never find but by discovering its own." (H. A., 283 ff.)

In reference to these and similar expressions of the alchemists, Hitchcock rightly calls our attention to Plotinus, who writes, for example (Enn., VI, 9, 10): "We must comprehend God with our whole being, so that we no longer have in us a single part that is not dependent upon God. Then we may see him and ourselves as it beseems us to see, in radiant beams, filled with spiritual light, or rather as pure light itself [notice this fullness of light] without weight, imponderable, become God or rather being God. Our life's flame is then kindled; but if we sink down into the world of sense, it is as if extinguished. . . . Whoever has thus seen himself will, then, when he looks, see himself as one who has become unified, or rather he will be united to himself as such a one and feel himself as such. Possibly one should not in this case speak of seeing. But as regards the seen, if we can indeed distinguish the seeing and the seen, and not rather have to describe both as one, which is, to be sure, a bold statement, then the seeing really does not see in this condition, nor does he differentiate two things, nor has he the idea of two things. He is, as it were, another; he ceases to be himself, he belongs no longer to himself; arriving there, he has ascended unto God and has become one with him, as a center that coincides with another center; the two coinciding things are here one, and only two when they are separated.

In this sense we speak of the soul's being another than God."

I recall also the passage in Amor Proximi where it is said that the earth will again be placed in Solis punctum. The center of the sun [God] is to be seen in the symbol ☉. We now understand the mystical difference between the hieroglyphs ☉ and ◯, between gold and alum. In order to express in the mercury symbol ☿ the accomplished union (represented by +) of ☉ and ☽, which takes place through the newly discovered central point, the symbol ☿ is also used.

I have mentioned the vedantic teachings, whose agreement with alchemy has also been noticed by Hitchcock. It takes emphatically the point of view of the " non-existence of a second." Multiplicity is appearance; the difference between the individual soul and the All Soul depends upon an error which we can overcome. The goal of salvation is the ascent into the universal spirit Brahma (in the nirvana of the Buddhists there is the same thought). Whoever has entered into the highest spirit, there is no longer any " other " for him. Brhadaranyaka-Upanishad, IV, 3: (23) " If he does not then [The man in the deep sleep (susupti),] see, he is yet seeing although he sees not, for there is no interruption of vision for the seeing, because he is imperishable; but there is no second beside him, no other different from him that he could see. (24.) If he does not smell, he is yet smelling although he smells not, for there is for the smelling [person] no interruption of smelling because he is imperishable; but there is no second thing beside him, no other thing different from him that he could smell. . . . (32.) He stands like water [i.e., so pure] seer alone and without a second . . . he whose world is Brahm. This is his highest goal, this is his highest fortune, this is his highest world, this is his highest joy; through a minute particle of only this joy the other creatures have their life."

If I compare the hermetic teachings on the one hand with the vedanta, and on the other with the Samkhya-Yoga, I do not lose sight of the fundamental antagonism of both — Vedanta is monistic, Samkhya is dualistic — but in appreciation of the doctrine of salvation which is common to both. That the mystic finds the same germ in both systems is shown by the Bhagavad-Gita. For him the theoretical difference is trivial, whether the materia is dissolved as mere illusion, when he has attained his mystic goal, or whether, as an eternal substance, it is as something overcome, simply withdrawn, never more to be seen. According to the Samkhya doctrine, too, the saved soul enters into its own being, and every connection with objects of knowledge ceases.

In Yogavasistha it is written: " So serene as would the light appear if all that is illumined, i.e., space, earth, ether, did not exist, such is the isolated state of the seer, of the pure self, when the threefold world, you and I, in brief, all that is visible, is gone. As the state of a mirror is, in which no reflection

falls, neither of statues nor of anything else — only
representing in itself the being [of the mirror]—
such is the isolation of the seer, who remains without
seeing, after the jumble of phenomena, I, you, the
world, etc., has vanished." (Garbe, Samkhya-Phil.,
p. 326.)

In the materia (prakri) of the Samkhya system reside the three qualities or constituents already familiar to us, Rajas, Tamas, and Sattva. Whoever
unmasks these as the play of qualities, raises himself
above the world impulses. For him, as he is freed
from antagonisms, the play ceases. When a soul
is satiated with the activity of matter and turns away
from it with disdain, then matter ceases its activity
for this soul with the thought, "I am discovered."
It has performed what it was destined to perform,
and withdraws from the soul that has attained the
highest goal, as a dancing girl stops dancing when
she has performed her task and the spectators have
enough. But in one respect matter is unlike the
dancing girl or actress; for while they repeat their
performance at request, matter "is tenderly disposed like a woman of good family," who, if she is
seen by a man, modestly does not display herself
again to his view. This last simile is facilitated in
the original texts by the fact that the Sanskrit for
soul and man has the same phonetic notation (pums,
purusa). (Garbe, l. c., pp. 165 ff.)

In comparing the common mystic content of Vedanta and Samkhya-Yoga with alchemy, I avoid the
difficulty involved in establishing a detailed con-

cordance of the hermetic philosophy with one or another system. An inquiry into this topic would result differently according to which hermetic authors we should particularly consider.

It is probably worthy of notice that the Yoga-Mystics, like the alchemists, are acquainted with the idea of the union of the sun and the moon. Two breath or life currents are to be united, one of which corresponds to the sun, the other to the moon. The expression Hathayoga (where hatha = mighty effort. Cf. Garbe, Samkhya and Yoga, p. 43) will also be interpreted so that Ha = sun, tha = moon, their union = the yoga leading to salvation. (Cf. Hatha-Yoga-Prad., p. 1.)

The union of two things, the sun with the moon, the soul with God, the seer with the seen, etc., is also taught by the image of the connection of man and woman. That is the mystic marriage (Hieros gamos), a universally widespread symbol of quite supreme importance. In alchemy the last process, i.e., according to the viewpoint of representation, the tincturing or the unification, is quite frequently represented in the guise of a marriage — sometimes of a king and a queen. We cannot interchange this final process with the initial one of introversion, which (as a seeking for the uterus for the purpose of a rebirth) is likewise readily conceived of as a sexual union. If the symbol of coitus was conceivable there, so here, too, the same symbol is appropriate for the representation of the definite union with the object longed for.

It is quite suggestive to associate the anagogic idea of the *Unio mystica,* precisely on account of the erotic allegory, with the primal motive of sexual union (with the mother) instead of with the wish to die, as I have done at another place.   It may be that the primal erotic power supplies something for the accomplishment of this last purpose; it may be that all powers must coöperate.   If I now still abide by my original exposition, this happens because it appears to me that the symbolism emphasizes the going over of the one into the other more than the attainment of the sexual goal; and even in the cases where the unio mystica is described as a sexual union.   We should not forget that the sexual gratification is to be regarded also as a kind of annihilation.   It is a condition of intoxication and of oblivion or perishing.   It is this side of the sexual procedure that the symbolism of the unio mystica particularly emphasizes.

Brhadaranyaka-Upanisad, IV, 3, 21:   ". . . For even as one embraced by a beloved woman has no consciousness of what is within or without, so the spirit, embraced by the most percipient self (prajena almana, i.e., the Brahm), has no knowledge of that which is external or internal.   That is its form of existence, in which it is characterized by stilled desire, even its own desire is without desire and separated from sorrow."   This passage treats of the deep sleep (susupti) which is regarded as a passing union with the highest spirit, and so, as essentially the same as the definitive *unificatio.*   Sleep is the brother of death. Susupti is, furthermore, conceived only as a preliminary; a German mystic would call it a foretaste of the definitive ascent into Brahm.

In the parable the unio mystica appears twice represented, once in that the king and queen are represented as the bridal couple, and the second time when the king, i.e., God, takes the wanderer up into his kingdom.

Th attainment of an inner harmony, of a serene peace, is what, as it seems to me, is most clearly brought out as the characteristic of the final unificatio — not merely by the Hindus or Neoplatonists, but also by the Christian mystics and by the alchemists.

Artephius is quoted by H. A., p. 86, as follows: ". . . This water [water of life] causes the dead body to vegetate, increase and spring forth, and to rise from death to life by being dissolved first, and then sublimed. And in doing this the body is converted into a spirit, and the spirit afterwards into a body; and then is effected the amity, the peace, the concord and the union of the contraries."

Similarly Ripley (H. A., p. 245): "This is the highest perfection to which any sublunary body can be brought, by which we know that God is one, for God is perfection; to which, whenever any creature arrives in its kind [according to its nature], it rejoiceth in unity, in which there is no division nor alterity, but peace and rest without contention."

The final character of the completed philosopher's stone makes it conceivable, that, as the hermetic masters say, it is made only once by a man and then not again. The Stone is an absolutely imperishable Good; but if it should be lost it is surely not the right stone.

I have now to offer some conjectures regarding further interpretations of the two and the three principles ☉ and ☽, namely ♇ ☿ ⊖. We are aware of a general difference. I add now first the remark of Hitchcock that the "two" things are to be regarded as an antithesis: *natura naturans* and *natura naturata*. We might intellectually conceive the ☿ (mercury) given by many writers at the beginning of the work as a double one, on the one hand as nature and on the other as our world picture. We cause it to work on our ♇ (sulphur), i.e., on our affectivity by which the ♇ is purified and dissolved, for it is compelled to adapt itself to the requirements of the world laws. But by this means a new world picture is produced, for the former had been influenced by the unclarified ♇ ; our affective life limits our intellectual. The new world picture or the newly gained ☿ we combine with our ♇ and so on, until finally after a gradual clarification nature and our world picture harmonize. Then there are no longer two mercuries but only one; and the sulphur, our completed subject, has become more or less a unity. Now we may advance to the unification of the two clarified things, which in this stage are called ☉ and ☽. Now subject and object are bound together and man enters, as is so wonderfully expressed in Chandogya-Upanisad, VIII, 13, as a being adapted into the unadapted (uncreated-primordial) world of Brahma. ☉ and ☽ may, to be sure, be conceived also as the love of God towards man and the love of man towards God. The different masters of the art are the same in different ways in that the one sees more the intellectual, the other the emotional. They describe different sides or aspects of the same process, for which we do not indeed possess appropriate concepts, and whose best form of expression is through symbols. The sign ☉ is then neither = subject nor love but just = ☉, i.e., a thing to which we may approximate nearest by a form of integration of all partial meanings. In view of the fact that ♄ and ☿ are contrasted at the beginning of the process also as body and soul, we can, by making ♄ = passions and ☿ = knowledge (reason) conceive the rest thus: ♄ is to be purified by an exalted ☿ (in distinction from the common ☿, called also " our " ☿), and so to be purified by a higher knowledge. From ♄ is developed (i.e., it unmasks itself to the initiated as) ☽, i.e., Maya, the object, that in its difference from the subject is mere illusion; and from ☿ comes ☉, the Brahm or subject, and now the *unio mystica* can take place. Another use of symbolism is the one by which we are able to concoct gold out of sulphur; from the affects we derive, through purification, love (toward God). The spirit ☿ exalts [raises] the antithesis ☉ and ☽ (soul and body) in such a way that finally it simply opposes itself as subject and object. (Cf. H. A., pp. 143 ff.)

Sometimes the making of gold is described as an amalgamation; from the raw material, ☉ is derived by an amalgamation with ☿ [quicksilver]. That naturally signifies the search for the Atman or highest spirit in man by means of contemplation, which belongs to ☿, the [act of] knowing.

With regard to the trinity ☉ ☽ ☿ :  The solar divinity [creating, impregnating] in man is △ that by its triangle moreover marks the fiery nature △ ; that which is comprised in the bodily nature, the terrestrial is ⊖ salt, which is also represented as a cube, like the element earth.  The two can be called ☉, anima, and ☽, corpus.  The celestial messenger who appears as a mediator for the antithesis is the conscience ☿, who has his constant influx from God, the real ☉, and is therefore a divine spirit.  We have then the triad Spiritus, anima, corpus [ ☿ ☉ ☽ ] or, because ☿ is to be regarded as a mediator, ☉ ☿ ☽.  The intervention of the ☿ effects the previously mentioned exaltation of ☉ and ☽ or of △ and ☿ (crude state) to ☉ and ☽.

In view of the difficulty of the mystic work that attempts to accomplish a sheerly superhuman task, it is not surprising that it cannot be finished in one attempt but requires time.  It necessitates great persistence.  In the life of the mystic the states of love and aspiration for God alternate with those of spiritual helplessness and barrenness.  (Horten, Myst., I, p. 9.)

Arabi sings in his ode on man's becoming godlike:
" [1] O thou ancient temple.  A light has arisen

for thee (you) that gleams in our hearts.   [2] To
thee I lament the wilderness that I have traversed,
and in which I have poured forth an unlimited flood
of tears.   [3] Neither at dawn nor at dusk do I
get repose.   From morning until evening I fare on
my way without ceasing.   [4] The camels go forth
on their journey at night; even if they have injured
their feet, they still hasten.   [5] These (mighty)
riding camels bore us to you (probably God) with
passionate longing, although they did not hope to
attain the goal. . . ."   The riding camels signify
the longing of the mystics for God.   " It seeks and
strives ceaselessly, although its powers are drained
by the difficulties of the search."   (Horten, l. c., p.
16 ff.)

Many degrees or stations are to be gone over on
the difficult way, yet zeal is to abide constant in all
circumstances.   [The idea of the ladder set up to
heaven, of steps, etc., is universal in religions.]   In
general seven such steps are distinguished.   In
Khunrath, e.g., the citadel of Pallas has seven steps.
Paracelsus (De Natura Rerum, VIII), following a
favorite custom, gives seven operations of the work.
". . . It is now necessary to know the degrees and
steps to transmutation, and how many they are.
These steps are then no more than seven.   Although
some count still more, it should not be so.   For the
most important steps are seven.   The further ones,
however, which might be reckoned as steps are comprised under the others, which are as follows: calcination [sublimation], dissolution, putrefaction, dis-

tillation, coagulation, and tincturing. Whoever passes over these seven steps and degrees comes to such a marvelous place, where he sees much mystery and attains the transmutation of all natural things." In the "Rosarium" of Johannes Daustenius [Chap. XVII] the seven steps are represented as follows: "And then the corpus [1] is a cause that the water is retained. The water [2] is the cause of preserving the oil so that it is not ignited on the fire, and the oil [3] is the cause of retaining the tincture, and the tincture [4] is a cause of the colors appearing, and the color [5] is a cause of showing the white, and the white [6] is a cause of keeping every volatile thing [7] from being no longer volatile." It amounts to the same thing when Bonaventura describes septem gradus contemplationis [seven steps of contemplation], and David of Augsburg [13th century] the "seven steps of prayer." Boehme recognizes 7 fountain spirits that constitute a certain gradation and in the yoga we also find 7 steps, which are described in the "Yoga Vasistha" (cf. Hath. Prad., pp. 2 ff). It may easily happen that the domination of the number 7 is to be derived from the infusion of the scientific doctrines (7 planets, 7 metals, 7 tones in the diatonic scale) and yet it may depend on an actual correspondence in the human psyche with nature — who can tell? Most significant is the connection of the 7 steps of development with the infusion of the nature myth in the alchemistic theories of "rotations." For the perfection of the Stone, rotations (i.e., cycles) are required by many authors, in which the materia (and so the soul) pass through the spheres of all the planets.    They have to be subjected successively to the domination (the regimen) of all seven planets. This is related to the ideas of those neoplatonists and gnostics according to which the soul must, on its way (anodos) to its heavenly home, i.e., to its celestial goal, pass through all the planetary spheres and through the animal cycle. (Cf. Bousset, Hauptpr. d. G., pp. 11 and 321.)    I observe, moreover, a thoroughly vivid representation of this very theme in the good old Mosheim, Ketzergesch., p. 89 ff.    Also in the life of the world, if it is completely lived, man passes through, according to the ideas of the old mystery teachings, the domination of the seven planets.

The anagogic meaning of rotation may be that of a collection of all available (seven in number) powers, in order finally to rise as a whole, to God.

More important, or at any rate more easily comprehensible, appears to me the trichotomy necessarily resulting from the course of the mystical work, a triplicate division that results in the three main phases, black, white, and red.    The black corresponds to introversion and to the first [mystic] death, the white to the " new earth," to freedom or innocence, red to love, which completes the work. This general arrangement does not prevent the symbols from being often confused by the alchemistic authors.    There are gradations between the main colors, all kinds of color play; in particular the so-called peacock's tail appears, which comes before the stable white to indicate the characteristic gayness of color of visionary experiences, and which marks the stage of introversion.

If one put into the center of vision, as goal of the work, the recovery of the harmonious state of the soul, one might express oneself about the three primary colors as follows: The paradisical state demands absolute freedom from conflict. We can attain this only by completely withdrawing from the external world whatever causes conflict in connection with the external world, so that there comes to pass with regard to it, a thorough-going indifference. This indifference is the black. The freedom from conflict (guiltlessness) in the now newly beginning life is the white. Previously, at the disintegration (rotting) of the material, one constituent part was removed and taken away. That is, the libido becomes free (love). It is gradually alloyed with the white material, which is dry (thirsty without thirst); sown in the white ground. Life is without conflict now drenched with love, red. This true red thus attained is permanent because it is produced [in contrast to mere instruction] from the heart of hearts, the roots of innermost feeling, which is subjected to no usury.

The mystical procedure can be realized in different degrees of intensity. The lowest degree is as a program with the mere result of a stimulation; the highest degree is a final transmutation of the psyche. If this goal is attained in life, we have acquired the terrestrial stone. In contrast, the celestial stone belongs with the eschatological concepts and the celestial tincture is the apokatastasis.

It is an interesting question whether the resolution of conflicts, with evasion of the process in the outer world, cannot be accomplished subjectively, by battles with symbols (personifications) and in symbols, thus amounting to an abbreviation of the process. Theoretically this is not impossible, for the conflicts do not indeed lie in the external world, but in our emotional disposition towards it; if we change this disposition by an inner development, the external world has a different value for the libido.

" The projection into the cosmic is the primal privilege of the libido, for it naturally enters into our perception through the gates of all the senses and apparently from without, and actually, in the form of the pleasure and pain qualities of perception. These, as we all know, we attribute without further deliberation to the object, and their cause, in spite of philosophical deliberation, we are continually inclined to look for in the object, while the object is often hopelessly innocent of it." (Jung, in Jb. ps. F., III, p. 222; with which compare the Freudian transference concept and Ferenczi's essay on " Introjektion und Übertragung," in Jb. ps. F., I, p. 422.) Jung calls attention to the frequently described immediate projection of the libido in love poetry, as in the following example from the Edda (H. Gering) :

" In Gymer's Courtyard I saw walking
The maiden, dear to me;
From the brightness of her arms glowed the heavens,
And all the eternal sea."

The mystic looks for the conflicts that he desires
to do away with, in man, the place where they really
exist. With this theoretical presumption the possible objection against all mysticism is averted, namely
that it is valueless because it rests merely upon imagined experiences, upon fanaticism. This objection,
though not to be overlooked, does not apply to mysticism, which accomplishes an actual ethical work
of enduring value — but to the other path that issues
from introversion, namely magic (not to mention
physical and spiritual suicide). This is nicely expressed, too, in an allegorical way by saying that
magically-made gold melts, as the story goes, or
turns into mud (i.e., the pretended value vanishes in
the face of actuality) while " our " alchemistic gold
is an everlasting good. The yoga doctrine, too, describes Siddhi (those imaginary wonders in which
the visionary loses himself) as transitory, only salvation alone, i.e., the mystical goal being imperishable.

As for the metaphysical import of the mystical
doctrine, I might maintain that the psychoanalytic
unmasking of the impelling powers cannot prejudice
its value. I do not venture at all upon this valu-
ation; but for the very purpose of bringing into
prominence a separate philosophical problem, I must

emphatically declare that if psychoanalysis makes it conceivable that we men, impelled by this and that " titanic " primal power, are necessitated to hit upon this or that idea, then even if it is made clear what causes us to light upon it, still nothing is as yet settled as to the value for knowledge of the thing discovered.

I am so far from wishing to derive a critique of the metaphysical import of the doctrine from psychoanalytic grounds alone, that I felt called upon to make claim only to a synthesis for the merely psychological understanding of mystic symbolism, a synthesis which I have attempted to block out as well as I was able in the present Part III of my book.

## Section III

## THE ROYAL ART

It has been mentioned that the work of perfecting mankind might be realized in different degrees of intensity, which might extend from complete living realization to mere sympathy without any clear comprehension. The psychic types in which the realization is achieved are, it may be said, identical.

These typical groups of symbols that the mystic [I draw a certain distinction between the mystes and the mystic. The latter is a mystes who makes a system of what he has realized.] produces as a functional expression of his subjective transformation, can be thought of as an educational method applied to arouse the same reactions in other men. In the group of symbols are contained more or less clearly the already mentioned elementary types as they are common to all men; they strike the same chords in all men. Symbolism is for this very reason the most universal language that can be conceived. It is also the only language that is adapted to the various degrees of intensity as well as to the different levels of the intro-determination of living experience without requiring therefore a different means of expression; for what it contains and works with are the elementary types themselves [or symbols which are as adequate as possible to them] which, as we have seen, represent a permanent element in the stream of change. This series of symbols is quite as useful to the neophyte as to the one who is near to perfection; every one will find in the symbols something that touches him closely; and what must be particularly emphasized is that the individual at every spiritual advance that he makes, will always find something new in the symbols already familiar to him, and therefore something to learn. To be sure, this new revelation is founded in himself; but there results for the uncritical mind (mythological level) the illusion that the symbols (e.g., those of the holy scripture) are endowed with a miraculous power which implies a divine revelation. [Cf. the concept of the origin of the symbol in my essay, Phant. u. Myth.] Because of a similar illusion, e.g., Jamblichus posits demons between gods and men, who make comprehensible to the latter the utterances of the gods; the demons, he thinks, are servants of the gods and execute their will. They make visible to men in works and words the invisible and inexpressible things of the gods; the formless they reveal in forms and they reveal in concepts what transcends all concepts. From the gods they receive all the good of which they are capable, partially or according to their nature, and share it again with the races that stand below them.

I said above, every one will find something appropriate to himself in the symbols, and I emphasized the great constancy of the types fast rooted in the

unconscious, types which impart to them a universal
validity. The divine is revealed " only objectively
different according to the disposition of the vessel
into which it falls, to one one way, another to another. To the rich poetical genius it is revealed pre-
eminently in the activity of his imagination; to the
philosophical understanding as the scheme of a harmonious system. It sinks into the depths of the soul
of the religious, and exalts the strong constructive
will like a divine power. And so the divine is honored differently by each one." (Ennemoser, Gesch.
d. M., p. 109.) " The spiritual element of the inheritance handed down by our fathers works out
vigorously in the once for all established style. . . .
On the dark background of the soul stand, as it were,
the magic symbols in definite types, and it requires
but an inner or outer touch [E.g., by religious ob-
servances.] to make them kindle and become active."
(Ib., p. 274.) " The unconscious is common to all
mankind in an infinitely greater degree than the content of the individual consciousness, for it is the
condensation of the historically average and oft-re-
peated." (Jung, Jb. ps. F., III, pp. 169 ff.)

Whoever allows the educative symbols to work
upon him, whether he sees only darkly the ethical
applications typified in them, or clearly perceives
them, or completely realizes them in himself, in any
case he will be able to enjoy a satisfying sense of
purification for his earnest endeavor in an ethical
direction. The just mentioned dim perception
(probably the most frequent case), does not exclude

the existence of very clear ideas in consciousness; the person in question generally considers his ideas, although they are only masks in front of the absolute ideal, as the ultimate sense of the symbol, thus accepting one degree of significance for the complete meaning. Every one approximates the ideal as he can; the absolute ideal through his ephemeral, but attainable ideal. The highest being speaks in the inexhaustible Bhagavad-Gita:

"More trouble have they who devote themselves to the invisible;

By physical beings the invisible goal is attained only with difficulty.

[XII, 5.]

" 'God is the all. Hard is it to find the noble man who recognizes this.

Those whom greed robs of knowledge go to other gods,

Cleave to many rulers—their own nature rules them,

And whatsoever divinity one strives to honor in belief,

I respect his belief and direct him to the right place.

If he strives in firm belief towards his divinity's favor and grace,

Then shall he in part get what he wants, for I gladly put good in his way.

Yet the result is but limited in the case of those of limited sensibility.

He finds the gods who honors them. Who honors me, attains to me."

[VII, 19–23.]

If above I derived the instructive group of symbols from a mystic, that is not to say that it must be precisely so. I brought out this case among possible

cases only for the reason that the mystic is the one
who carries out most strenuously the ethical work of
purification, and under such conditions as are most
favorable to a suggestive group of symbols, and in
particular those rich in characteristic types.  Bear
in mind the founders of religion.  (They do not always have to be individuals — schools, myths.)
There are, however, others than the religiously inspired natures, who are preëminently endowed to
produce suggestive symbol groups with anagogic
value; the artists.  I suspect that it would prove
that the purifying (cathartic) action of a work of
art is the greater the more strongly the anagogic
symbolism (the groups of types that carry it) is developed, or in other words, the more they express
a tendency to a broadening of the personality.  This
tendency, to which belong the motives of the denial
of the selfish will (father figure) of the love that is
connected with sacrifice (incest motive, regeneration) of the devotion to an ideal (longing for
death), etc., is manifested in the artist as also in the
devout observer of the work of art in his very devotion to it.  Being lost in a work of art appears to me
essentially related both to introversion and to the
unio mystica.

I have already spoken of the creations of the
myth forming imagination and its anagogic import.
In alchemy, to which I wish now to return, the mythical and the individual images meet in the most vivid
way, without destroying each other.

In regard to the high ethical aspirations of alchemy, we understand that as a mystic art it preserves those attributes of a royal art which it seems to have had at first merely as gold making and magic. In fact what art may more justly be called royal than that of the perfection of mankind, that art which turns the dependent into the independent, the slave into a master? The freeing of the will in the mystic (and in every ethical) process has already, I believe, been commented on enough to be comprehensible. And the power of rule that has been extolled as a magical effect of the Philosopher's Stone lies in the harmonizing of the individual will with that of the world or with God's will. In the new birth — so remarks Jane Leade casually — we acquire a magic power; this occurs " through faith, that is, through the harmony of our will with the divine will. For faith puts the world in our power, inasmuch as the harmony of our will with the divine has the result of making everything ours or obedient to us. The will of the soul, when it accords wholly with the divine, is no longer a naked will lacking its raiment, power, but brings with it an invincible omnipotence."

To-day, too, there is a royal art. Freemasonry bears this name. Not only the name but its ethical ideal connects it with the spirit of the old alchemy. This statement will probably be contradicted and meet the same denial as did once the ideas of Kernning [J. Krebs], although I think I am on different ground from that of this poetic but, in my eyes, all too uncritical author. Keep in mind the historical treatment mentioned in Part I, Section 4, and furthermore do not forget the psychological basis of our present modes of viewing things.

[If I wished to compare the ethical aims of both in general terms, I should run the risk of unduly expatiating on what is easily understood. Robert Fischer describes freemasonry as a society of men who have set themselves the severe task of a wise life and labor as the most difficult task, of self-knowledge, self-mastery and self-improvement,— tasks that are not finished in this life but only through death prepare us for the stage where the true consummation begins. These beautiful and straightforward words could just as well stand in an alchemistic discussion on the terrestrial and celestial. But this will suffice.]

And now permit me to present the following portrayals by Jane Leade [English mystic of the 17th century. She belonged to the philadelphian society founded by Pordage.] which I reproduce here with a few words of comment, and take them as an illustration of the beautiful spiritual union of the serious hermetic with the new royal art. The reader can draw his own conclusions. The passages are taken from Leade's " Garten-Brunn " (L. G. B.). References to Wirth are to the " Symbolisme Hermétique " (W. S. H.) of this modern author.

This mystic who is sunk in deep meditation on the noble Stone of divine Wisdom, has a vision of Sophia (Wisdom) at which she is startled. " Soon came the voice and said: Behold I am God's ever-

lasting handmaid of wisdom, whom thou hast sought.
I am now here to unseal for thee the treasures of the
deepest wisdom of God, and to be to thee even that
which Rebecca was to her son Jacob, namely, a true,
natural mother.   For from my body and womb shalt
thou be born, conceived and reborn."   (L. G. B.,
I, p. 14.)

Leade is much rejoiced that the "morning star
from on high" has sought her, and secludes herself
for the following days to await further developments.   She has still more visions of the crowned
queen of heaven and was asked whether she had the
desire to be taken up into the celestial company.   She
proves herself of constantly devoted will and from
this time wisdom speaks to her as an inner illumination.   (L. G. B., I, p. 15 ff.)

[Retirement is a precondition of introversion and
of withdrawal into oneself.   The uninitiated who
is to be admitted is, to use the language of alchemy,
the subjectum, in whom the process of purification
is to be perfected.   The alchemists put the subjectum
into a narrow vessel so as to be hermetically sealed
from the outer world.   There it is subjected to
putrefaction as in a grave.   Introversion leads into
the depths of one's very heart.   "Where were you
formed?"   "In my heart [or inner man]."
"Where after this?"   "In the Way to the Lodge."
"What determined you . . .?"   "My own free
and unconstrained will."   The uninitiated are recommended to take counsel seriously with regard to
their important resolution.   "Why are you . . .?"

"Because I was in darkness and desired light."
The death symbol in the sch. K. is later to be considered. I can naturally go into a few only of the
analogies. The informed reader will largely increase the number of parallels very easily.]

Jane Leade seeks in the spirit for the key that can
open the entrance into the great secret that lies deep
hidden within her. Her effort to reach the holy city
is great but at first ineffectual. [One is not admitted
without further effort.] She wanders around the
city and finds no entrance. [Way to the Lodge —
"Why have they not led you the nearest way to the
Lodge?" "In order to acquaint me with the difficulties and troubles that one must first overcome before one finds the way of virtue."] She is apprehen-
sive that she must, lacking the wonderful key, now
grope all her days in darkness . . . never find the
gate. "While I, now overpowered with fear and
horror at all this, was plunged [Symbols and processes in the sch. K. Roll of the terrible Br. It is
probably well founded psychologically, a fact that I
should like to emphasize in opposition to Fischer,
Kat. Erl., I, on Question 7.] into a deep silence and
stillness, the word of wisdom itself was revealed to
me and said: 'O deeply searching spirit, be not surprised that you have not realized your hopes for so
long a time. So far you have been with many others
caught in a great error, yet as you know and are
sorry for your error, I will apprize you what sort of
a key it must be. . . . And although this wonderful
Key of Wisdom is a free gift, it will yet come to be of

high value to you, O searching spirit, when once you
obtain it.' " Nevertheless wisdom goes about and
looks for those who deserve it, [Nothing being made
of nothing, the point of departure of the philosophic
work is the finding and choice of the subject. The
material to work upon, say the alchemists, is quite
common and is met with everywhere. It is necessary only to know how to distinguish it, and that is
where all the difficulty lies. We continually experience it in masonry, for we often initiate the profane,
whom we should have rejected had we been sufficiently clear sighted. Not all material is good to
make a mercury. The work can succeed only if we
succeed in finding a suitable subject; so masonry
makes many inquiries before admitting a candidate
to the tests. (W. S. H., p. 87.)] She does this so
as to write herself on the inner walls of their hearts
and in each and every one meet their thoughts which
wait upon her laws and counsel, [Obedience of ap-
prentices. The laws of wisdom are meant.] and
brings a kingdom with it which will be well worth
sacrificing everything for. [Laying aside of all
metal. The newly admitted brother is " through his
unclothing (which probably belongs here) to represent mankind symbolically, as he comes from the
hand of nature, and to remind us that the freemason, in order to be continuously mindful of the
fulfillment of his duty must be able to rid himself of
all fortuitous externalities. See Note H at end.]
But the greatest and most distinguished master work,
says wisdom, consists alone in your keeping your

spirit disciplined and learned, and making it a skilled worker or artist, to give it knowledge of what material, as well as in what number, weight and measure [Surveying, geometry.] to make this pure key, which [material] is the bright pure divinity in the number three, the mighty in truth. . . . It is distinguished as a surpassingly mighty glory and lordliness which sits in a circle of heaven within the hearts of men. [The connection of circle (doubly significant) and heart is interesting. As is well known the circle is placed on the bleeding left breast. In the old English ritual the touching with the point of an instrument (sword or the like) is proved " Because the left breast is nearest to the heart, so that it may be so much the more a prick in my conscience as it then pricked my skin."] It does with the plumbline of its power measure the temple and inner court with those who worship there. [The line in connection with the temple: the " binding " on a carpet; an image of the curtain string in the holy of holies in Solomon's temple. " Just as this ribbon holds and closes the curtain, so an indissoluble bond unites and holds together all free and accepted Masonic brothers (also those who worship therein)." This is wisdom's key [Surveying, Geometry.] which will make our hands drop with sweet smelling myrrh upon the handles of the lock (Solomon's Song, v, 5). When now I opened your secret gate with this key, my soul failed within me and I had no strength in me, the sun of reason and the moon of my extended senses were confounded and vanished. I knew nothing by myself of

the active properties of nature and the creature.
[" What have you seen as . . .? " " Nothing that
reason can grasp."]   The wheel of motion stood
still and something else was moved by a central fire,
so that I felt myself turned into a bright flame.
Whereupon this word came to me:   This is nothing
else than the gate of my everlasting depths; can you
stay in this fiery region, which is wisdom's dwelling
and abode, in which it meets holy remote spirits and
gives them a fiery principle?   For if thou canst take
heed such that thou comest hither at its order; then
no secret shall be kept from thee.   So far I have been
permitted to approach the entrance to your house,
where I must still stay until I hear further from you
what is to be done.   (L. G. B., I, pp. 17–19.)

[As we hear it is therefore right to keep the
spirit corrected and disciplined.   " Why came ye . . .
to subdue my passions — to subject my will . . ."
We see two triads.   A divine three (3 great lights),
and then sun, moon and central fire, which second
three can be called the lesser lights, as the " M. v.
St." appears as a central fire.   If we remember that
the didactic voice proceeds, according to this symbolism, from a fire or light (Wisdom), this light is
identical with the M. v. St. in the function, and it is
determined by exactly that.   The central fire is
naturally also the blazing star.   This stands on the
tapis between sun and moon and it is designed to
illuminate the innermost space of the temple.   From
alchemy we are well acquainted with $\odot$, $\mathrm{D}$ and an
intermediate and mediating light, namely $\mathary$.   This

light can be also symbolized as ⚵ . To the alchemistic point of view correspond quite closely the three great lights of the Freiburg Ritual, God, man and St. John's light. This (the ☿ ) is the intelligence and the talent in men which creates all science and shows us the truth. It is " the only authority that the freemason has to recognize unconditionally, namely, the divine ordinance in his own heart, the celestial fire in his ego."]

Several weeks later Leade hears again the voice of wisdom. It said to her: " Separate thyself and withdraw from thy animal sensuous life, it is too coarse. I cannot appear till that is completely lost and vanished." [The alchemistic separation (separatio) and the masonic taking off of parts of their clothing. I have already made the most necessary remarks about it. We have to be freed from the things which, as in the eclectic ritual " much retard the soaring of the spirit and chain man to the earth." It has an expressly programmatical meaning (anticipating a later phase) when, e.g., the system of the Grand Lodge goes back, for the deprivation from the metal, to " the temple of Solomon that was built of fully prepared stones, just as they were brought, etc.," so that no metal work was needed.] A short time after Leade is again driven to search after the secret being. Wisdom requires it to know itself apart from its creature existence. " Whereupon I was surrounded by gently burning flames that consumed and burned all thistles, thorns and accursed fluxes [the " superfluities " of alchemy] which would

put themselves forward. . . . Therefore Wisdom
let her voice be heard and said: ' O thou troubled
spirit, I am now come to tell thee what was required
of thee, as I have not refrained from saying to thee,
even at the beginning of my discourse with thee,
what it would cost thee to attain this key. [First
programmatically shown. The actual process will
then follow.] I say to thee, God requires a sacrifice
of thee. Understand me then, thou hast an earthly
element that has spread and covered [Like a garment.] thee, and consequently has got the upper
hand and mastery of thee; these thrones and powers
[king or father figure] must, however, be over-
thrown and their place found no more. Thou hast
deeply mourned that thou hadst to do without the
ever near communion or union with God thy Creator,
[Only the masters sit near the sun.] but be not surprised at that. The cause lies here in complete ex-
tinction because you are not yet deceased and dead
completely in the mystical manner. [Complete ex-
tinction first results in the third degree.] This is
the first baptism that you must experience, but ah,
how many have rushed into this too abruptly, because they have not given their earthly selfishness a
single mortal blow right to the heart. [The circle
or sword placed on the left breast alludes merely to
the process of the clearing of conscience. Here the
whole ego is not yet annihilated.] So I recommend
to thee my flaming sword. Be courageous and let it
achieve complete execution in the field of nature
[The weed in the field is exterminated where, as

Jane Leade frequently says, ears of corn are to grow.] or banish completely all young or old, and turn from life toward death whatever in you does not bear my mark and name that is my image.'" [From this the psychological sense of the countersign is recognized. In connection with the field we are reminded especially of Shibboleth (Judges XIII, 6: The Ephraimites who could not speak it had to die). Leade often mentions the Ephraimites. Directly pertinent to the above passage is, of course, Revelations, passim.] (L. G. B., I, pp. 21 ff.) The earthly is, as it were, to be sacrified to God as a burnt offering or melted away in a fiery furnace, in a vessel of the purest metal. [Probably it will not be superfluous to remark that in the Bible the first worker in all kinds of metal was Tubal Cain, whose name is a password.] Jane Leade finds " the conditions or circumstances which thou [Wisdom] requirest of me to be very hard; especially do I find myself still dwelling in the offspring of a mortal shadow, where whole millions of spirits tempt me and employ all their ability and strength to hinder and hold me back from the high and noble exaltation and aspiration, [The seductive and restraining voices in the circuitous way or on the way to the Lodge according to the eclectic ritual. The band corresponds to the mortal shadow.] while I, alone and seeing the receptacle and fire before me, stood in thought about it and pondered the matter, and was willing, like Isaac, to ask, But where is the lamb? [The apron is of lamb's fleece.] She [Sophia =

Wisdom] answered my unspoken question with these words:   You yourself must be the paschal lamb that shall be slain.   Thereupon I was instructed to say or to beg:   Then give yet this life pulse a stroke so that it may completely return.   And as I stretched out my neck, so to speak, to the love flaming sword, I felt that a separation or beheading had taken place. [Note the baring of the neck, the guttural and its meaning given according to the content of the old form of oath.   The fate of the traitor befalls the man who is slain at this point; he has been a traitor to the inner true man.   It is here the place to bear in mind the descending scale of marks (guttural, pectoral, stomachal).   Man is to be transmuted on rectangular principles, or in the language of alchemy, is to be tinctured with the divine tincture.   This tincturing seizes first the most spiritual and advances steadily until the whole man is transformed.   The trichotomy corresponds to the Platonic (and alchemistic) tripartite division of the powers of the soul.   Plato distinguishes the reasoning soul, which he places in the head, the intellectual in the breast, and the affective in the abdomen.   The entire soul, even the vegetative, is to be illuminated by the highest light.   If we assume that it is more than a pretty picture, Staudenmaier's view becomes of interest, namely that man may have an extraordinary spiritual perfection in bestowing consciousness through practice upon the centers that ordinarily work vegetatively without consciousness.   In this way he gains power over a whole army of working powers that otherwise es-

cape him. Staudenmaier's own experiences teach
that all the dangers of introversion are connected
with such a training, and it may easily happen that
we are defeated by the spirits that we invoke, instead
of becoming their masters. The absolute mastery of
the rational ego is, however, evidently the foundation of the ethical work of perfection. Kenning's
doctrine is related to the theories of Staudenmaier.]
Oh, how sweet and pleasant it is to perceive the life
blood flowing into the fountain of the same divinity
from which it came." Whereupon wisdom opened
more of her secrets to her.    (L. G. B., I, p. 24.)

It may be that this is the most suitable place to
mention another series of visions (apropos of the
building of the tabernacle, L. G. B., I, pp. 24 ff.) :
" It [the holy ark] is an impregnable fortress and
tower, so go thou not out [so says Wisdom], but
bind thyself and ally thyself here as a disciple, to
hold out to the end, then thou wilt be learned in the
lofty spiritual art of the everlasting mystery, and
be instructed how this incomparable composition or
medicine of the healing elixir and balsam of life is
prepared. Above all thou must enter a bond of
silence and vow to reveal it to no one outside of your
fellow learners, who are called together near and
with you, to work at this very art. [I hardly need
to mention the duty under oath, but will only call
attention to the group of the three virtues of the
newly entering: attentiveness, silence, fidelity.] Further thou must completely bide the definite time and
year of it, in all fidelity and patience indefatigable,

until thou succeedest in making this oil as well, and preserve it in the beautiful snow-white alabaster box of consummate nature, and art as fit and perfect as thy instructress."

I continue in the earlier series of forms. Jane Leade is required by wisdom to follow her. " But all of a sudden I was surprised by a mighty enemy, who pressed me hard while he accused and complained that I was breaking the laws of nature, to which I was still bound because I had an external body, for whose elemental wants I must take reasonable care, . . . as all my neighbors in the world did, who were under the rule of the grand monarch of the [worldly calculating] reason, under whose scepter everything must mortify what lives in the sensual animal life. . . ." [The man who lives only for the satisfaction of physical needs cannot serve our purpose. . . . There is a higher life than that to which millions are chained like an animal. To this higher life the Master is to devote himself, and to it he is metaphorically initiated in the admission. Common nature, the prince of this world, strives against these requirements.] " Yes," says the prince of the earthly life, " how wilt thou turn aside from my laws and throw thy brother's yoke from thy neck? " Leade turns to her mother, Wisdom, who promises her to take God's advice how the enemy could be driven away. The proof should be that they were traitors to the crown, to honor, and to the lordship of the lamb; they would soon be handed over to justice. (L. G. B., I, pp. 27 ff.)    [Cf. on the one hand, in connection with "accused and complained," think of the murderer of the royal architect. As this is the inner man, both belong together. The "prince of this world" turns the tables on his accusation; psychologically quite justifiably.]

After various exhortations Jane Leade receives from Wisdom a book which she, Wisdom, must read from, "in order to explain to you one letter after another, [Spelling.], especially you do not yet know the number which makes up your new name. And as long as you do not see that, what kind of right and title can you advance for the rest of the entire mastery that is developed there?" (L. G. B., I, p. 36.) It refers to a transmutation of the man, which cannot happen all at once; "so highly important a change, that it could not take place without a passing through many distant degrees." (L. G. B., II, p. 78.)

We come to a section that is inscribed, "The Magic Journeys." [Probably I shall hardly need now to refer to the meaning of the journey.] It contains all the other phases of the mystical work. "During my spiritual journey to the land of all blessed abundance, a magic outline of it was placed before my eyes, while I was brought to a door which was so low and narrow that I could enter it only by creeping through on my knees, so that it also required great effort and trouble. [Obstacle of the door.] And so I was led farther till, after some time, I came to another door, which was indeed narrow enough but somewhat more comfortable to go through than the first. As I thus proceeded, I came finally to a door that had two valves, one of which opened itself, and was quite right in height and width for my size, and also admitted me to a place of which I could find neither beginning nor end. And I said, ' What am I doing here alone? ' Whereupon my invisible guide who had led me through these three doors or gates replied that still others would come after me, when they should hear that there was anywhere so great a country that was to be possessed by new inhabitants, and that should be filled and blessed also with all kinds of goods." (L. G. B., I, p. 40.) [The three gates refer not merely to the three degrees, but have still for themselves another analogue in the initiation. In the old English system the aspirant knocks, because the door offers him a resistance, on the backs of the three officials. They are, as it were, the spiritual doors of the brotherhood. The resistance, and how it is gradually presented in Leade's description, is readily understood psychologically; the nature of the aspirant is the more adapted the further he advances on his work.]

"This idea and apparition and the account and explanation following thereupon were very powerful; so that I entered into the thoughts of it ever deeper, . . . so that I . . . also might perceive the explanation and meaning of the gates. For although my spirit saw naught but an infinite spaciousness [compare previous pages] I perceived and felt [Infinite spread of the lodge in accordance with the examination.] still the blowing of so fragrant and refreshing a breeze, as if all kinds of flowers actually stood blooming there.    [Does the question as to the mason's wind belong here psychologically?   In any case the pleasant breeze comes from the east.   Jane Leade often describes her flower garden as oriental. Psychologically and mythologically the breeze has the value of a spermatic symbol.   Anagogically it is concerned with the bestowing of a power or (to retain the procreation metaphor) the impregnation with such a power.]    Therefore this word was revealed and spoken to me:   ' This is space and place where the love realm is to arise and become verdant with its natural inhabitants, who have laid aside their crass self-love [selfishness] and left it behind them, as it might not come here; even as it is the one which makes the entrance so narrow and crowded. . . .'   Hereupon I saw in my spirit unexpectedly different persons, modified out of measure in their bodies, and they were so highly versed in this mystery that they breathed forth such a spirit from them that they could give being and existence to everything that they willed and desired.    At times they spread golden tents and went in and out of them, at other times in places that appeared to be quite waste and desolate they made wonderful plants and trees to grow up, which actually offered their perfect fruit that appeared in a bright golden radiance; of which it was related that they were the magical nourishment and food on which the inhabitants of this land were supposed to live."   [We may also say the masters of the art cultivate an uncultivated people, and provide spiritual nourishment at the drawing table.] "And although at my first entrance here it seemed that I saw nothing, yet I did see after a few moments this whole spacious place filled with spirits of so high a degree that they attracted me at once. Thereupon they set me divers philosophical questions [Catechizing.] which I did not understand. So one of them in a very friendly manner offered to instruct and teach me; said further that he would teach me the secret of their art. . . . Accordingly he brought me into a magnificent tent, and requested me to wait there so that I might advance into the pure acts or works of faith, because I would succeed thereby in becoming an adept in this high philosophy. Now when, thereupon, Wisdom herself appeared to me, I asked her who it was that had set me the philosophical questions? Whereupon she answered me that they were the old and last living worthies, and they were the believing holy ones, taught by her in her inward and outward divine magic stone; and that the time was coming in which she desired to make new artists and poets in this theosophic wisdom, who should give a form to the things that had been so odiously disgraced and lay under a cloud of contempt, ignorance and disgrace. Especially, no other way besides this could be found than that the deep mine, in which the treasure had lain hidden so long, should be broken open and unsealed. [The lost that must be found again is called in freemasonry the master word. The master wandering has the object of seeking what was lost there

[in the East] and has [partly] been found again.]
Hereupon the apostle John [Well beloved] spoke
to me, to whom the secret was well known, and who
was the person who had spoken so kindly to me before, with these words:  'Just as a natural stone, so
is there also a spiritual stone which is the root and
the foundation of all that the sons of art have
brought visibly into being and into the light.   And
just as the external is corporeal, and consists in work
of the hands, and consumes a great amount of time
before it can be brought to perfection, so also is the
internal elaborated from degree to degree. . . .'
Therefore I begged and asked the angel John in what
manner I might go to work, to work out the same?"
The "angel John" accorded her the permission.
Just as a furnace is used for a chemical preparation,
so also a furnace is necessary for the preparation of
the spiritual Philosopher's Stone.   This outer furnace is, however, the corporeal man, in which "the
fire seeds of pure divinity itself are kindled by the
essence of the soul, when he finds for it a hallowed
and properly prepared vessel.   The materia in
which one must labor or work is the divine salt, which
is placed in a pure clear crystalline glass, the pure
spirit.   Further shouldst thou know, that this divine
salt is concealed in all men."   (L. G. B., I, pp.
40–43.)

    Here I must insert the discussion of the salt (also
salt stone) and its effect.   We must understand
clearly that the salt stone of this symbolism is the
same as the cubical stone of the masonics.   That the

salt is hieroglyphically represented by a cube, I have
already shown.   The concept stone is the parting of
the ways for two symbol groups of similar meaning,
both of which Jane Leade uses.   The one group
is the chemical preparation, as the angel John
just now described it; the other is the treatment of
the stone as a building stone (which is to be dressed,
etc.) found in other passages from Jane Leade,
namely, in connection with the building of a temple,
a sanctuary, the New Jerusalem.   Important use is
made of men as building stones preëminently in her
" Revelation of Revelations."   This one passage
from L. G. B. (II, p. 138) may be quoted:   " Who
will now blow this trumpet of mine that they may
break loose from their iron yokes and bonds and
come hither, so that they may become worthy to be
built in as well-cut pillars for the temple of wisdom."
The quadrangular form is several times mentioned
also.   Jane Leade is quite right when she says that
the divine salt, the cubic stone, lies concealed in all
men; the unprepared man is the crude stone and in
him lies to be developed (potentially not actually)
the cubic.   In the preparation of the stone, the alchemistic as well as the building stone, it depends on
the clearing away of the disturbing elements, not on
ornamentation.   The purification (rectificatio, puri-
ficatio, etc.) of the alchemistic stone exactly corresponds to the working over the raw stone with the
pick.   Crystallization produces the regular form;
fixation, the density.   The projection corresponds to
the employment in the building of the temple (which

appears infrequently in symbolism). Probably the most appropriate place for these passages (L. G. B., I, pp. 131 ff.) is in connection with this mention of building. " Only have faith, so I will go before you and reveal my name and show you the foundation of it [the city], wherefrom your strength increases and your victorious power shall be known. But who must be your architect to instruct you in this foundation work of yours but this Wisdom who was with the great God Jehovah from eternity, who gave you existence and being from the breath of the eternal Will? Therefore thus and in such manner the motivating power of the will must result and proceed. . . . Come therefore to me and I will show you where all these foundation stones lie. Look and see the material of the treasure in the circumference of your new earth. . . . Here you might spy out or find this foundation, [Cf. what was said previously about the new earth], for which purpose will be given to you the golden measuring line, or the plummet of my spirit." [The master stands on the pillar of strength. Jehovah was the last master word.]

We stopped where the angel John says we should know that the divine salt is hidden in all men. It goes on: " but it has lost its power and savor, and such is the principle of light that includes all other principles, because man, although quite unknown to himself, is an abstract and concept in brief of all worlds. Therefore he may find in himself all that he seeks; only it cannot happen before the salt alone, which has lain as dead, has been again raised to life through Christ the Freestone (who calcines the black to a jasper brilliance and to a beautiful whiteness). This is the true theosophic medicine, which indeed gradually, or little by little, works out of itself, from itself, and into itself, even as a grain of wheat which when it is sown does, by the coöperation of the sun and the outer planets, forms itself into a body. Only one has to watch and pay attention so that no birds of prey come and pick it up [Cf. Figure 3, p. 199] before it comes to its maturity and full time. For just such a state [as with the grain of wheat] exists in the case of the gold stone, which lies hidden in the foundation of nature, is nourished by the warm fiery influence of the divine sun and through the moist seeds of the spiritual Luna [sperma luna] is watered, which makes it grow through the inner penetration and union of the planetary powers of the higher order, which draw the weaker and lower into themselves, impregnate and swallow them. Whereby the mastery is obtained over all that is astral and elemental.   In this manner the beloved John revealed to me the nature of the royal stone, as it was revealed to him in the island of Patmos (there by him was brought forth what he possessed in the spirit).   And he told me further concerning this: that where the universal or general love is born in any one, such would be the true signature and token that this seraphic stone would there be formed and take to itself a bodily shape." (L. G. B., I, p. 44).

[Here we meet clearly the trinity ☉ ☿ ☽, sun,

moon, and as an outgrowth of both the ⚥ gold stone, the Philosopher's Stone, which unites in itself the ☉ and ☽ or which is the same △ and ▽. It is therefore not at all a mistake to see in the ▽ a union of action and reaction. The G must be conceived in the anagogic sense, as the genesis of the Philosopher's Stone or as regeneration.] In L. G. B., I, p. 147, I find also this remarkable passage: "The word of Jesus was revealed to me in the following manner: O you that wait upon Jerusalem. Through what gate have you entered? And what have you seen here that you are so desirous of living here? Have you not been taken in by the fire flaming eye? [The eye is the flaming star. In L. G. B., I, p. 196, is found the representation of a face that is equivalent to the eye. A moon is added to it. The eye ☉ is, as it were, the sun to this moon.] so that you intend not to go out again from here, till you get another heart [The pectoral learns who approaches to the flaming star.] which never could be completely changed? . . . O then be therefore wise, and await your nuptial spirit [Genesis] and the garment of the power unfailing. [i. i. d. St.] No one can ever get that outside of this treasure city, for in this Zion all must be born anew. . . ." [Oswald Wirth regards the alchemistic concept Rebis as the expression of the perfect degree of community. "The initiated, who becomes in some way androgynous, because he unites the virile energy with the feminine sensitiveness, is represented in alchemy by the Rebis [from res bina, the double thing]. This substance, at once male and female, is a mercury ☿ animated by its sulphur 🜍 and transformed by this act into Azoth ☿, i.e., into this quintessence of the elements [fifth essence] of which the flaming star is the symbol. It should be noted that this star is always placed in such a way that it receives the double radiation of the male sun ☉ and the female moon ☾; its light is thus of a bisexual nature, androgynous or hermaphrodite. The Rebis corresponds otherwise to the matter prepared by the final work, otherwise called the journeyman who has been made worthy to be raised to the mastery." (W. S. H., p. 99.)]

But to return to Jane Leade's magical journey. "Hereupon I was moved (because I well knew and was certain that this heavenly stone already had its birth and growth in me) [Rebirth = the cubical stone's change from potentiality to actuality] with great frankness to ask whether my external furnace [her own body] would keep so long, and not perish [die] before the stone would have attained its perfection. Whereupon this dear saint [John] said to me in answer: Worry and trouble not yourself about this but be only patient in hope; for the true philosophic tree is grown and in a fair way to produce ripe fruit." (L. G. B., I, pp. 44 ff.) The preparation of the stone is now described by John according to the well known outlines. For " said Wisdom and the apostle John to me: Henceforth you shall be brought to the old worthy heroes of the faith who have [The masters too.] effected projec-

tion with the stone [= the work of transmutation].
And after I was brought there I saw the patriarchs
or arch fathers and all the great philosophers, who
had been taught by God himself both in the earlier
and in the later times.  After that I was led into a
darkness and gloom, which was of itself changed by
a magic power into a clear silver light."  (L. G. B.,
I, p. 46.)  Several other allegories follow for the
changing activity, as described already (L. G. B., I,
p. 41).  John explains that all the wonders were
accomplished with the stone of wisdom and that whoever has worked out this stone in himself is marked
as one sealed [Sigillum, Hermetes, Sealing with the
Trowel, mark of salvation, Mark Mason.] of God
with the power from above.

A further communication of the preceding vision
[sc. Magic Journey] gives the following additional
information:  "The Word came to me and said:
The love bond between God and thee must not be
loosened but tightly knotted.  Meanwhile the spirit
is the only eternal substance and property in which
thou must labor and toil.  That it may then cling
to you so fast and strongly that it may draw thee
quite closely to it and may establish thee within the
circle of the immeasurable love; from which enmity
is sundered, and the curse of the elements is separated and wholly taken away.  O go in, go, I say,
into it, for this is the infinite space, that thou hast
seen, and which is to be found inside the third door.
[Does this need any explanation?]  This invisible
love bond will perfect thee through the first gate,

which is so narrow and low, and therefore also through the other two gates; in case that thou wilt yield everything in thee completely in all its length and breadth so that it may be able quickly to raise thee. For, dear one, what is to return thee so mightily to the desired enjoyment of all abundance and good as the love of God? Therefore be strong and courageous in love, in going through these divers gates, and fear not any attack of the adversary till thou hast entered this hallowed country and art wedded therein to thy beloved."

A complaint that was made of Wisdom by her pilgrim: " Meanwhile as I lay in my deep struggle, came there a spirit of prayer down, who made an earnest supplication and unutterable sighing, rise towards heaven, [The lamentation at the grave of the Master.] which as I felt most clearly, penetrated and broke through the gate of the eternal profound, so that my spirit had an entrance to the secret chamber of pure godhead, wherein I had audience and complete freedom to pour out my lamentations and show my wounds and tell who had pierced me. For each and every hand was against me, let fly their stinging arrows at me, and burdened and oppressed still more that which hung already, dropping blood, upon the cross, and cried and said, Crucify, crucify her, make her really feel death in the dying. . . . I was in violent birth travail. All woes and onsets, however, made a greater opening for the birth of life, and gave me an entrance into the holy place, wherein first I heard the eternal tones.

And then after this, as I gained the strength to be in a pleasant quiet, I was in a clear water, [Tears.], in which no mud nor any refuse arose; also no implement was lifted to any work there, nor was any noise nor uproar heard." [Just as in the building of Solomon's temple.] (L. G. B., I, p. 48.)

Now Leade hears the comforting voice of the "Bridegroom" (the unio mystica) which brings to her view the perfection she has striven for, and commands her to touch no unclean spirits of this world. [Gloves.] Only what is detached from sin may come near him. The bridegroom is answered by Leade's soul-spirit: "Lord, how can this be done? For although I have had a great longing towards this ministration [the holy service] that I might be ever near thee, the spirit of this world [See previous.] has made claim to this shell or body of mine, and says that I have not yet stepped out the bounds and sphere of his dominion. The external man is encompassed by hunger and thirst, heat and cold [antitheses of the Hindu philosophy], which are wont to entangle his external senses in such things as are external, in such a way that no one can live in such pure abstraction and seclusion, until he is relieved of and freed from all care for the external body. This is what I bewailed with tears, and expressly asked God whether it was not possible for the eternal mind and spirit to supply all necessaries for the bodily part without aid of the spirit of reason, who is king in that realm where malediction rules?" [In other words, whether it was not possible in the living life to be released from contradiction (as it is called in the Bhagavad-Gita), to quite tear away the bonds of animal sensuous being, and definitively allow the eternal principles to be active. The question is whether the terrestrial stone is, in its complete perfection, on the whole possible, whether the ethical ideal in absolute purity can be pragmatically realized.]

" Whereupon after a short blocking and stilling of my external senses, I received this answer [of the Bridegroom]; that this could not be until a complete death of the body of sin was suffered, showing me that which is written in the 6th verse of the seventh chapter of Romans, that after that was perished and dead, wherein we were held, we should serve God in newness of spirit." (L. G. B., I, pp. 50 ff.)

[Here we have then the requirement to become wholly dead to the realm of sin, in order to be able to rise fully to the ethical ideal. The question whether this is possible in life remains open, to be sure. In symbolism this mystical death and the union with the highest spirit was represented symbolically in the highest degree of freemasonry. The representative of the Highest is the Master degree of the M. v. St. and he fills the dead, as it were, with his life, as the raising takes place (H in H, F against F, K against K, etc.), like the reviving of the child by Elijah (I Kings XVIII, 21). As for the necessary decay of the body before the raising (" The skin leaves," etc.) let us quote the passage, L. G. B., I, pp. 271 ff.: [the divine word speaks] " Know . . . that I have not left thee without a potent and rich talent which lies in thine own keeping, although deep hidden and covered with a threefold covering (Exod., XXXIX, 34, Num. IV, 5, 6), which must be removed before thou canst see this costly garment. The first covering is the coarse dark appearance of this earthly realm . . . the second is the fast-binding [directed upon the mundane] reason . . . the third is the baser natural senses. . . . Provided that thou thoroughly determine with the firm resolution to break through these three obstacles, thou wilt come to the golden mass. . . . While it is given to ye then to know where the treasure really lies [Seeking out of the grave. The three murderers, who have hidden the corpse, are these very " three obstacles."] and you have my spirit on this, which will not alone seek for it but will with the hand of its strength strongly coöperate with you; [To revive the mystical dead.] so resolve as united in the spirit . . . to break through that and to break it up, which lies as a covering over this princely being. . . . Pray and do not only wait [no idle mysticism] but struggle and work until you have released, set free and liberated this power hidden in a prison; which may be exalted upon the throne of empire, since in truth my spirit as well as yours has hitherto not been displaced from its kingdom except by force and unrighteousness." [With reference to the raising (cf. L. G. B., III, pp. 87, 91); in that place three degrees of mystical development are described in the similitude of three altars. Under the last altar, which is built of quadrangular stones,

in which one can see his own face as in mirrors, lies
the life trampled to death, which will again be
wakened.]

"Knowest thou not, I was asked, that the law
of sin has the mastery as long as it [the body with
its selfish tendencies] lives? So that the spirit
clearly bore witness and gave me to understand that
nothing could make me worthy of this marriage with
the Lamb [unio mystica] except an absolute death,
since he wedded only the maidenly spirit, to be one
flesh with him, [H in H, F against F, etc.] and by
so doing changed it into his own pure manhood.
[Humanity.] And this is the generation or birth
to an actively self-sufficient being, which rises out of
the old one. For just as the grain of wheat perishes
or dies in the earth and comes into a new life, just so
it goes also with the arising and the growing up of
the new creation, which in truth is Christ our life,
whose appearance will put an end to the sins in us.
For, dear one, what has brought on the curse, care,
trouble, misery, weaknesses, which press and torture
the poor man in this fallen state of his but the departing from God? And as long as he is in this
condition, he is a debtor to sin and under its dominion; which subjects him to all affliction and misery, which are wont to follow the footsteps of those
who live in the elemental flesh. Now without doubt
it is good and joyful tidings to hear of a possibility
of drawing out and putting off this body of sins; and
in truth the prophet who has arisen in me has prophesied that such a day was at hand. Be dismayed at

it, ye that are the wounders and despisers of this
grace; which I see is now near being revealed, for the
bridal garments are being prepared. . . . [Cf. the
end of the parable.] O Wisdom, the preparation
and ordering of the bridal garments is given in
charge to you alone, which shall be of divers colors,
with which the king's daughter, [Analogue of the
king's son, the improved son figure of the parable.]
who is entrusted to thy teaching and instruction, may
be distinguished from all others, and known [as re-
deemed]." (L. G. B., I, pp. 51 ff., wherewith the
magic journey is ended.)

Thus there is a confident tone, a hope in that which
loses itself in the infinite. But Leade suspects that
it is an unattainable ideal and knows what regulative
import it has: "Ah, who up to this hour has trav-
eled so far, and what are all our realized gifts until
we have reached this goal [union with the Divinity].
Can our plummet even sound it and explore in the
deep abyss, the matchless wonder of the immeasurable being? And because the revolving wheel of my
spirit has found no rest in all that it has seen, known,
possessed and enjoyed, it stretched its errant senses
continuously towards what was still held back, and
kept, by the strong rock of omnipotence; to struggle
towards which with a fresh attack I resolutely determined, and would be sent away with nothing less
than the kingdom and the ruling power of the Holy
Ghost." (L. G. B., I, p. 87.)

In a parallel between the old and new royal art,
I cannot overlook the French masonic writer Oswald

Wirth, who has worked in the same province. I
agree with him in general; although much of his
method of interpretation seems to me too arbitrary.
I have already called attention to several passages
from W. S. H. on the preparation of the subject
[i.e., the uninitiated]. I will endeavor to outline
the contents of the rest of the work according to the
ideas of Wirth.

Having given up himself, the Subjectum is overcome in the philosophic egg [preparation chamber,
i.e., sch. K.] by sadness and suffering. His strength
ebbs away, the decomposition begins; the subtle is
separated from the coarse. [Smaragdine tablet.]
That is the first phase of the air test. After descending to the center of the earth [Visita interiora terrae,
etc.— Smaragdine tablet, 6, 8.] where the roots of
all individuality meet, the spirit rises up again
[Smaragdine tablet, 10.] released from the caput
mortuum, which is blacked on the floor of the hermetic receptacle. The residuum is represented by
the cast-off raiment of the novice. Laboriously now,
he toils forward in the darkness; the heights draw
him on; escaping hell he will attain to heaven. His
ascent up the holy mountain is hindered by a violent
storm; he is thrown into the depths by the tempest:
a symbol of circulation in the closed vessel of the
alchemists, which vessel corresponds to the protected
lodge. During the circulation the volatile parts rise
and fall again like rain, which is symbolized by the
tears upon the walls. To be sure, it is not here that
the neophyte is subjected to the water test, and if a

confusion is possible on this point it comes from the
fact that all the operations of the great work go on
in one vessel, while the masonic initiation is completed in a suite of different rooms, so that the symbolic series here suffers disintegration.   The circu-
lating water, which soaks into the pores of the
earthly parts of the subject, purifies it more and more,
so that it goes from gray through a series of colors
(peacock's tail) to white.   In this stage the material
corresponds to the wise man who knows how to re-
sist all seduction.   Yet we are not to be satisfied with
this negative virtue; the fire test (we should remember that the four tests by the elements were to be
found in the parable also) is still to be gone through,
the calcination, which burns everything combustible.
After the calcination there is a perfectly purified
salt ( $\ominus$ ) of absolute transparence.   So long as the
novice has not attained this moral clearness, the light
cannot be vouchsafed to him.   In brief, in the first
degree, the main thing is the comprehensive purification.   The salt layers must be made crystal clear,
that surround the inner sulphur $\varphi$ like a crust and
hinder it from its free radiation.   Sulphur is to be
regarded as a symbol of the expansive power, as individual initiative, as will.   Mercury stands opposite
to it as woman does to man, as that which goes to
the subject from without, or as absolute receptivity.
Salt is midway between both; in it the equilibrium
between $\varphi$ and $\ddot{\varphi}$ is found.   It is a symbol of what
appears as the stable being of man.   In the first de-
gree the purification of the salt is worked out for

the release of the sulphur.   The red column J corresponds to the red sulphur, by which symbolically the
novices get their reward.   For the rest, the first degree is satisfied with getting the novices to see the
universal light (the blazing star).   (W. S. H., pp.
88–92.)

The fire test takes place in the second degree.
The fiery sulphur must be worked out or rather sent
out and used for work.   The field of activity of the
member proportions itself, as it were, according to
the expansion or range of its sulphurous radiation.
At this time the member enters into a relation of
such intensified activity with the world that the intellectual grasp [which corresponds to the ☿ = prin-
cipal] acquires from it a new illumination [blazing
star], and breaks away for a connection of the will,
which was at first merely individual, with the collectivity.   To me at least that appears to be the
sense of the figurative but not quite clear exposition
of W. S. H., pp. 952–962, which I have, for the
sake of exactness, given in the original text.   [See
Appendix, Note I.]   As soon as the crude stone is
cut and polished we have no longer to work inward
but outward.   What we are to accomplish so creatively would be insignificant if we did not know the
secret of borrowing power from a power that apparently lies without us.   Where do these mysterious powers work if not at the pillar B, whose name
means: i. i. d. St.?   In the north directed on the
contrary towards the moon, whose soft feminine
light it reflects, it corresponds to ☿, which unceas-

ingly flows towards all being, in order to support its
central fire, △. The exaltation of the latter leads
to the fire test, the idea of which Wirth seems to
take in strictly occult form, in the manner of Eliphas
Levi. Finally, a circulation takes place, in that the
individual will seeks like a magnet to draw the divine
will, always falls down again, rises, however, and so
on in cycles, till both meet in the " philosophical fire."
It is the cycle of which we read in the Smaragdine
Tablet. The incombustible essence that comes forth
from the fire test is the phœnix (a figure much used
by the alchemists). The member has the task of
changing himself into the phœnix. Not only △
belongs to the work, however, but also the act must
be guided by intelligence; activity and receptivity
must complement each other. Therefore the member has to know both pillars thoroughly. And
therefore he becomes also the already mentioned
androgynous material, Rebis. That is only to be
attained when the elemental propensities are overcome, therefore the figure Rebis is represented as
standing on the dragon. (W. S. H., pp. 96–101.)

What will the master do now? He will identify
himself with the Master Builder of all worlds, in
order to work in him and through him. When any
one says that that is mysticism, he is not wrong.
Being developed on the three successive ways of pur-
gatio, illuminatio, and unio, this mysticism is no less
logical than the religious mysticism that with its
mortifications, if it were only rightly understood,
would accomplish the same purpose. Mortification

is, as the word itself says, the endeavor for a certain kind of death.    Twice is the mason enjoined to die; at the beginning in the preparation room and at the end at the final initiation into the inmost chamber. The second death corresponds to the perfection of the grand mastery.    It signifies the complete sacrifice of his personality, the renunciation of every personal desire.    It is the effacement of that radical egotism that caused the fall of Adam, in that he dragged down spirituality into corporeality.    The narrow pusillanimous ego melts into nothing before the high impersonal self, symbolized by Hiram.    The mythical sins of the eternal universal human Adam are thus expatiated.    The architect of the temple is to the Grand Master Builder of All Worlds (G. B. a. W.) just what in Christian symbolism the Word become flesh is to the Eternal Father.    In order to carry on the work of the universal structure with advantage the Master must enter into the closest union of the will with God.    No longer a slave in anything he is the more a master of all, the more his will works in harmony with the one that rules the universe.    " Placed between the abstract and the concrete, between the creative intelligence and the objective creation, man thus conceived, appears like the mediator par excellence, or the veritable Demiurge of the gnostics."    Yet it is not enough that he gets light from its original source, he must also be bound by endless activity to those whom he is to lead.    The necessary bond is sympathy, love.    " The master must make himself loved and he can only succeed by himself loving with all the warmth of a generosity extending even to absolute devotion, even to the sacrifice of himself." The pelican [We are already acquainted with this hermetic bird.] is the hieroglyph for this loving sacrifice without which every effort remains vain. (W. S. H., p. 105.)

The master's degree, this necessarily last degree, corresponds to an ideal that is set us as a task: we must strive towards it even if its realization is beyond our powers. Our temple will never be finished, and no one expects to see the true eternal Hiram arise in himself. (W. S. H., p. 94.)

We find also in Wirth, how the work is divided into three main steps, which begin with the purifying, turn towards the inner soul, and end with the death-resembling Unio Mystica; here we find, too, in the last degree the unattainable ideal, which like a star in heaven shall give a sure course to the voyage of our life. The viewing of the exalted anagogic conception as a perspective vanishing point, makes allowance for the possible errors of superposition in the anagogic aspect of the elementary types.

The tripartite division, which we meet in the great work, shows the frequently doubted inner qualification of the three degrees of freemasonry. As they answer a need, they have again prevailed, although they were not existent in the masonic form of the royal art at the beginning (about two centuries ago); I say " again," because similar needs have already earlier produced similar forms. (Cf. L. Keller's writings.) Whether we consider ethical education in general or the intensive (introversion) form of it, mysticism, we have in either case a process of development, and degrees are necessary to express it symbolically. The effort, appearing from time to time, to multiply the degrees has been justified. We can divide what is divided into three sections into seven also (7 operations in alchemy, 7 levels of contemplation, 7 ordinations, etc.), although it is not really needed. But the idea of abolishing the three degrees can only arise from a misapprehension of the value of the existing symbolism. That masonry is a union of equal rights is not affected by the presence of the degrees, provided that their symbolic significance is not overstepped. The degrees form a constituent part of the symbolic custom itself and like it are to be intangible.

The symbols of all the lofty spiritual religious communities, for which the royal art presents itself as a paradigm or exemplar, put before us, as it were, types of truth. Single facts which the symbols may signify (or that could be read into the symbols) are not the most important, but rather the totality of all these meanings. The totality (which can be acquired only by a sort of integration) is something inexpressible; and if it also succeeded in expressing this inexpressible, the words of the expression would be incomprehensible to any finite spirit, as the individual facts are.

The symbols are the unchangeable, the individual meanings are the variegated and the changeable. [As for the masonic symbolism in particular, I am in agreement with Robert Fischer (Kat. Erl., III, fin.). "Freemasonry rests on symbols and ceremonies; in that lies its superior title to continued existence. They are created for eternal verities and peculiarly adapted thereto; they are fitted to every grade of culture, indeed to every time, and do not fall like other products of the time, a sacrifice to time itself. . . . Therefore a complete abolition of our symbols can meet with assent as little as an enfeeblement of them can be desired. Much more must we strive in order that a clear understanding may sift out the abstract, corresponding to our spiritual eye, from the concrete necessary for our physical eye, so that the combined pictures shall be resolved in the simple fundamental truths. By this means the symbols attain life and motion and cannot be put down for things that decay with time."] Therefore the symbols should never be changed in favor of a particular meaning, which becomes the fashion (or be brought closer to it over and above the given relation). What is to be maintained through variations of meaning, is not the meanings but the symbols themselves.

To each person symbols represent his own truth. To every one they speak a different language. No one exhausts them. Every one seeks his ideal chiefly in the unknown. It matters not so much what ideal he seeks, but only that he does seek one. Effort itself, not the object of effort, forms the basis of development. No seeker begins his journey with full knowledge of the goal. Only after much circulation in the philosophic egg and only after much passing through the prism of colors does that light dawn which gives us the faint intimation of the outline of the prototype of all lesser ideals.   Whoever desires hope of a successful issue to this progress must not forget a certain gentle fire that must operate from the beginning to the end, namely Love.

> **Whom these teachings cheer not,**
> **Deserves not to be a man.**

# NOTES

Note A (80). I put here not merely those comparisons of motives which are alike at the beginning, but also those that are important for our further consideration. My rendering of them is partly abridged. Signs of similarity are, as Stucken explains, not employed to express an absolute congruence, but predominantly in the sense of " belongs with " or " or is the alternate of." Stucken's comparison I, A, goes: Moses in the ark = spark of fire in the ark = Pandora's books = Eve's apple; I, B: Moses in the ark = the exposed = the fatherless = the persecuted = the deluge hero [the one floating in the ark]. II, A: Eve's apple = Moses in the ark = Onan's seed = fire = soma = draught of knowledge, etc. III, B: Tearing open of the womb = decapitation or dismemberment = exposure = separation of the first parents. IV, B: The dismembered [man or woman] = the rejuvenated = the reborn [m. or w.]. VI, A: Potiphar motive = separation of first parents = Onan motive. VII, A: The wicked stepmother = Potiphar's wife = man eater. VII, B: Flight from the " man eater = flight from Potiphar's wife = flight from the wicked stepmother = separation of the first parents = magic flight. IX, A: The first parents = magic flight. IX, A: The killed ram = Thor's ram = Thyestes' meal = soma. XIII, A: The exposed = the persecuted = the dismembered child = the slain ram — the helpful animal. XIX: The Uriah letter = the changed letter = word violence [curse = blessing]. XX: Scapegoat = ark. XXVIII: Wrestling match = rape of women = rape of soma = opening of the chest [opening of the hole] = rape of the garments [of the bathing swan ladies]. XXIX: Castration = tearing asunder [consuming] of the mother's body = the final conflagration = the deluge. XXXIII, A: Dragonfight = wrestling match = winning of the offered king's daughter = rape of the women = rape of fire = deluge. XL, A: Incest motive = Potiphar motive. XL, B: Incest = violation of a [moral] prohibition. XL, C: Seducer [male or female] to incest = "man-eater." XLIV, A: The father who rejects the daughter = "man eater." XLIV, B: Separation of the first parents = refusal of the daughter [refusal of the "king's daughter" promised to the dragon fighter] = substitution. XLV, A: Sodomy = substitution = rape = parthenogenesis = marriage of mortal with the immortal = seduction = adultery = incest = love = embraces of the first parents = wrestling match. Otherwise is marriage of mortal with the immortal = incest = separation of the first parents. (SAM Book 5.)

Note B (128). That the ideas of gold and offal lie very near each other is shown in numerous forms and variations in myth, fairy tale and popular superstition. I mention above all the figure of the ducat or gold-dropper which has probably been attenuated from a superstition to a joke, and around it are gathered such expressions as "he has gold like muck," "he must have a gold dropper at his house"; then the description of bloody hemorrhoids as golden veins; the fabulous animals that produce as excrement gold and precious things. Here belong also the golden ass [K. H. M., No. 36], which at the word "Bricklebrit" begins "to spew gold from before and behind," or [Pentam., No. 1], at the command, "arre cacaure," gives forth gold, pearls and diamonds as a priceless diarrhea. [Arre is a word of encouragement like our get-up; cacuare is derived naturally from cacare, kacken = to cack, perhaps with an echo of aurum, oro, gold.] It occurs frequently in sagas that animal dung, e.g., horse manure, is changed into gold as, inversely, gold sent by evil spirits is easily turned [again] into dung. Gold is, in the ancient Babylonian way of thinking, which passes over into many myths, muck of hell or the under world. If a man buried a treasure so that no one should find it, he does well to plant a cactus on the covered [treasure] as a guardian of the gold, according to an old magical custom. An attenuation of the comparison dung = gold seems to be coal = gold. In Stucken we find the comparison excrement = Rheingold = sperm [S. A. M., p. 262] and connected with it [pp. 266 ff.] a mass of material mythologically connected .with it. I mention the similar parallels derived from dream analysis (Stekel, Spr. d. Tr., passim), further in particular the psychologically interesting contributions of Freud on "Anal character (Kl. Schr., pp. 132 ff.) like Rank's contribution. (Jb. ps. F., IV, pp. 55 ff.)

Note C (280). According to Jung it is a characteristic of the totality of the sun myth which relates that the "fundamental basis of the 'incestuous' desire is not equivalent to cohabitation, but to the peculiar idea of becoming a child again, to return to the parents' protection, to get back into the mother again in order to be born again by the mother. On the way to this goal stands incest, however, i.e., necessity in some way to get back into the uterus again. One of the simplest ways was to fructify the mother and procreate oneself again. Here the prohibition against incest steps in, so now the sun myths and rebirth myths teem with all possible proposals as to how one could encompass incest. A very significant way of encompassing it is to change the mother into another being or rejuvenate her, in order to make her vanish after the resulting birth [respective propagation], i.e., to cause her to change herself back. It is not incestuous cohabitation that is sought, but rebirth, to

which one might attain quickest by cohabitation.    This, however, is not the only way, although perhaps the original one."
(Jung, Jb. ps. F., IV, pp. 266 ff.)    In another place it is
said:    The separation of the son from the mother signifies the
separation of man from the pairing consciousness of animals,
from the lack of individual consciousness characteristic of
infantile archaic thought.    " First by the force of the incest
prohibition could a self-conscious individual be produced,
who had before been, thoughtlessly one with the genus, and
only so first could the idea of the individual and conclusive
death be rendered possible.    So came, as it were, death into
the world through Adam's sin.    The neurotic who cannot
leave his mother has good reason; fear of death holds him
there.    It appears that there is no concept and no word
strong enough to express the meaning of this conflict.
Whole religions are built to give value to the magnitude
of this conflict.    This struggle for expression, enduring
thousands of years, cannot have the source of its power in
the condition which is quite too narrowly conceived by the
common idea of incest; much more apparently must we
conceive the law that expresses itself first and last as a
prohibition against incest as a compulsion toward domestication, and describe the religious system as an institution that
most of all takes up the cultural aims of the not immediately
serviceable impulsive powers of the animal nature, organizes
them and gradually makes them capable of sublimated employment."    (Jb. ps. F., IV, pp. 314 ff.)

Note D (274).    Jung divides the libido into two
currents lengthwise, one directed forward, the other backward:    " As the normal libido is like a constant stream,
which pours its waters into the world of reality, so the resistance, dynamically regarded, is not like a rock raised in the
river bed, which is flowed over and around by the stream,

but like a back current flowing towards the source instead of towards the mouth. A part of the soul probably wants the external object, another part, however, prefers to return to the subjective world, whither the airy and easily built palaces of the phantasy beckon. We could assume this duality of human will, for which Bleuler from his psychiatric standpoint has coined the word ambitendency, as something everywhere and always existing, and recall that even the most primitive of all motor impulses are already contradictory as where, e.g., in the act of extension, the flexor muscles are innervated." (Jb. ps. F., IV, p. 218.)

Note E (279). Of the wonderful abilities that pass current as fruits of the yoga practice, the eight grand powers [Maha-siddhi] are generally mentioned: 1. To make oneself small or invisible [animan], 2, 3. to acquire the uttermost lightness or heaviness [laghiman, gariman], 4. to increase to the size of a monster and to reach everything even the most distant, as e.g., to the moon with the tips of the fingers [mahiman or prapti], 5. unobstructed fulfillment of all wishes, e.g., the wish to sink into the earth as into water and to emerge again [prakamya], 6. perfect control over the body and the internal organs [isitva], 7. the ability to change the course of nature [vasitva], and 8. by mere act of will to place oneself anywhere [yatra kamavasayitva]. Besides these eight marvelous powers many others might be named, which are partly included in the above; such an exaltation of sensitiveness that the most remote and imperceptible images, the happenings in other worlds on planets and stars, as also the goings on in one's own interior and in other men's are perceived by the senses; the knowledge of the past and the future, of previous existences and of the hour of death; understanding the language of animals, the ability to summon the dead, etc. These miraculous powers, however, suffer from the disadvantages of being transitory, like everything else won by man through his merit — with the exception of salvation. (Garbe, Samkhya and Yoga, p. 46.)

Note F (305). Jung (Jb., III, p. 171) refers to Maeterlinck's "inconscient supérieur" (in "La Sagesse et la Destinée") as a prospective potentiality of subliminal combinations. He comments on it as follows: "I shall not be spared the reproach of mysticism. Perhaps the matter should none the less be pondered: doubtless the unconscious contains the psychological combinations that do not reach the liminal value of consciousness. Analysis resolves these combinations into their historical determinants for that is one of the essential tasks of analysis, i.e., to render powerless by disconnecting them, the obsessions of the complexes that are concurrent with the purposeful conduct of life. History is ignorant of two kinds of things: what is hidden in the past and what is hidden in the future. Both are probably to be attained with a certain measure of probability, the former as a postulate, the latter as a historical prognosis. In so far as to-morrow is contained in to-day, and all the warp of the future already laid, a deepened knowledge of the present could make possible a more or less wide-reaching and sure prognosis of the future. If we transfer this reasoning, as Kant has already done, to the psychologic, the following things must result; just as memory traces, which have demonstrably become subliminal, are still accessible to the unconscious, so also are certain very fine subliminal combinations showing a forward tendency, which are of the greatest possible significance for future occurrences in so far as the latter are conditioned by our psychology. But just as the science of history troubles itself little about the future combinations which are rather the object of politics, just so little are the psychological combinations the object of the analysis, but would be rather the object of an infinitely

refined psychological synthesis, which should know how to
follow the natural currents of the libido. We cannot do
this, but probably the unconscious can, for the process takes
place there, and it appears as if from time to time in certain
cases significant fragments of this work, at least in dreams,
come to light, whence came the prophetic interpretation of
dreams long claimed by superstition. The aversion of the
exact [sciences] of to-day against that sort of thought-process
which is hardly to be called phantastic is only an overcom-
pensation of the thousands of years old but all too great inclination of man to believe in soothsaying."

Note G (317). The umbilical region plays no small
part as a localization point for the first inner sensations
in mystic introversion practices. The accounts of the
Hindu Yoga doctrine harmonize with the experiences of the
omphalopsychites. Staudenmaier thinks that he has, in his
investigations into magic, which partly terminated in the
calling up of extremely significant hallucinations, observed
that realistic heavenly or religious hallucinations take place
only if the " specific " nerve complexes [of the vegetative
system] are stimulated as far down as the peripheral tracts
in the region of the small intestine. (Magie als exp.
Naturw., p. 123.) Many visionary authors know how to
relate marvels of power to the region of the stomach and of
the solar plexus. In an essay on the seat of the soul, J. B.
van Helmont assures us that there is a stronger feeling in
the upper orifice of the stomach than in the eye itself, etc.;
that the solar plexus is the most essential organ of the soul.
He recounts the following experience. In order to make
an experiment on poisonous herbs he made a preparation of
the root of aconite [Aconitum napellus] and only tasted it
with the tip of his tongue without swallowing any of it.
" Immediately," he says, " my skin seemed to be constricted
as with a bandage, and soon after, there occurred an extraor-

dinary thing, the like of which I had never experienced before. I noticed with astonishment that I felt, perceived and thought no longer with my head, but in the region of my stomach, as if knowledge had taken its seat in the stomach. Amazed at this unusual phenomenon, I questioned myself and examined myself carefully. I merely convinced myself that my power of perception was now much stronger and more comprehensive. The spiritual clearness was coupled with great pleasure. I did not sleep nor dream, I was still temperate and my health perfect. I was at times in raptures, but they had nothing in common with the fact of feeling with the stomach, which excluded all coöperation with the head. Meantime my joy was interrupted by the anxiety that this might even bring on some derangement. Only my belief in God and my resignation to his will soon destroyed this fear. This condition lasted two hours, after which I had several attacks of giddiness. I have since often tried to taste of aconite, but I could not get the same result." (Van Helmont, Ortus Medic., p. 171, tr. Ennemoser, Gesch. d. Mag., p. 913.)

Note H (381). For the old as for the new royal art the material is man, as man freed from all framework. "Not man of the conventional social life, but man as the equally entitled and equally obligated being of divine creation, enters the temple of humanity with the obligation always to remain conscious of his duty and to put aside everything that comes up to hinder the fulfillment of the highest duty." (R. Fischer.) Compare with this what Hitchcock says of the material of the Philosopher's Stone: "Although men are of diverse dispositions . . . yet the alchemists insist . . . that all the nations of men are of one blood, that is, of one nature; and that character in man, by which he is one nature, it is the special object of alchemy to bring into life and action, by means of which, if it could uni-

versally prevail, mankind would be constituted into a brotherhood." (H. A., pp. 48 ff.) [The tests] . . . "begin
with the stripping of the metals. Now alchemy recom-
mends, once the propitious matter is seen, carefully examined
and recognized, to clean it externally for the purpose of freeing it of every foreign body that could adhere accidentally to
its surface. The matter, in fine, should be reduced to itself.
Now it is an absolutely analogous matter that the candidate
is called to strip himself of everything that he possesses arti-
ficially; both it and he ought to be reduced strictly to themselves. In this state of primitive innocence, of philosophic
candor retrieved, the subject is imprisoned in a narrow retreat where no external light can penetrate. This is the
chamber of meditation which corresponds to the matras of
the alchemist, to his philosophic egg hermetically sealed. The
uninitiated finds there a dark tomb where he must voluntarily die to his former existence. By decomposing the in-
teguments that are opposed to the true expansion of the germ
of individuality, this symbolic death precedes the birth of the
new being who is to be initiated." (W. S. H., pp. 87 ff.)

Note I (411). As to the Chamber of the Com-
panion hung in red, it represents the sphere of action of our
individuality, measured by the extent of our sulphurous radiation. This radiation produces a kind of refractive [re-
fringent] medium, which refracts the surrounding diffused
light [ ☿ is meant] to concentrate it on the spiritual nucleus
of the subject. Such is the mechanism of the illumination,
by which those benefit who have seen the blazing star shine.
Every being bears in himself this mysterious star, but too
often in the condition of a dim spark hardly perceptible. It
is the philosophic child, the immanent Logos or the Christ
incarnate, which legend represents as born obscurely in the
midst of the filth of a cave serving as a stable. The initiation becomes the vestal of this inner fire △ ; archetype or

principle of all individuality.　She knows how to care for it as long as it is brooded in the ashes.　Then she devotes herself to nourishing it judiciously, to render it keen for the moment when it finally should overcome the obstacles that imprison it and seek to hold it in isolation.　It means, as a matter of fact, that the Son is put en rapport with the external ☿, or in other words, that the individual enters into communion with the collectivity from which he comes.

# BIBLIOGRAPHY

## OLD. (Before 1800.)

Agrippa ab Nettesheym, Henricus Cornelius, Opera.

Alchemisten, Griechische, v. Berthelot, Hoefer (moderne).

"Allgemeine (und General-) Reformation der gantzen (weiten) Welt" vide "Fama."

"Amor proximi" ("Das Buch Amor Proximi Geflossen aus dem Oehl der Göttlichen Barmhertzigkeit . . .") Ans tag-licht gegeben per Anonymum. Franckfurt und Leipzig, 1746.

(Andreae, Valentin) Anonym, Chymische Hochzeit Christiani Rosencreutz. Anno 1459. Gedruckt zuerst zu Strassburg bey Lazari Zetzners seel. Erben MDCXVI. Regenspurg, MDCCLXXXI (eigentlich Berlin, bei Nicolai).

Arabi v. Horten (mod.).

Arnaldus de Villa Nova, Chymische Schrifften. Aus dem Latein übersetzet durch Joh. Hoppodamum. Frankfurt und Hamburg, 1683.

Bādarāyana v. Deussen (mod.).

Basile, Giambattista v. mod.

"Bauer" ("Der grosse und der kleine Bauer"). Zwey Philosophische und Chymische Tractate. Leipzig, 1744.

Beaumont, Iohann, Historisch-Physiologisch- und Theologischer Tractat von Geistern. Aus der Englischen Sprache in die Teutsche mit Fleiss übersetzt von Theodor Arnold. Halle im Magdeburgischen, 1721.

Bekker, Balthasar, Die Bezauberte Welt. Amsterdam, 1693.

Bernhard, Graf von der Mark und Tervis, Abhandlung von der Natur des (philosophischen) Eyes. Hildesheim, 1780.

" Bhagavad-Gîtā " v. Schlegel, Schroeder (mod.).

Boehme, Jacob, Schriften. Gesammtausgabe von Johan Georg Gichtel. 2 Bde. 1715.

Bonaventura, Opera.

Booz, Ada Mah (Dr. Adam Michael Birkholz), Die sieben heiligen Grundsäulen der Ewigkeit und Zeit. Leipzig, 1783.

" Confessio " (der Fraternität R. C.) v. " Fama."

Dastyn, John (Ioannes Daustenius), Rosarium, Visio. Vide " Sieben-Gestirn."

Eisenmenger, Johann Andreä, Entdecktes Judenthum. 2 Tle. Königsberg, 1711.

Eleazar, R. Abraham, R. Abrahami Eleazaris Uraltes Chymisches Werk. In II Theilen zum öffentlichen Druck befördert durch Julium Gervasium Schwartzburgicum. Erfurt, MDCCXXXV.

" Fama Fraternitatis oder Entdeckung der Bruderschafft dess löblichen Ordens dess Rosen-Creutzes, Beneben der Confession Oder Bekanntnuss derselben Fraternitet . . . Sampt einem Discurs von allgemeiner Reformation der gantzen Welt." Franckfurt am Mayn, M.DC.XV.

Fictuld, Hermann, Des . . . Chymisch-Philosophischen Probier-Steins Erste Classe, in welcher der wahren und ächten Adeptorum . . . Schrifften nach ihrem innerlichen Gehalt und Werth vorgestellt . . . worden. Dritte Auflage. Dresden, 1784.

" Figuren der Rosenkreuzer." (" Geheime — aus dem 16-ten und 17-ten Jahrhundert.") Drei Hefte. Altona, 1785ff.— Titel der einzelnen Hefte: I. AVREVM SECVLVM REDIVIVVM. Henricus Madathanus, Theosophus. II. Ein güldener Tractat vom Philosophischen Steine. Von einem noch Lebenden, doch vngenanten Philosopho . . . beschrieben. Anno M.-DC.XXV. III. Einfältig A B C Büchlein für junge Schüler so sich täglich fleissig üben in der Schule des H. Geistes . . . Von einem Bruder der Fraternitet CHRISTI des Rosenkreuzes P. F.— Auf dem Titelblatt der ersten beiden Hefte heisst es: " Aus einem alten Mscpt zum erstenmal ans Licht gestellt." In jedem Heft folgt dem Text eine Reihe von farbigen Tafeln.

Flamellus, Nicolaus, Chymische Werke. In das Teutsche übersetzt von J. L. M. C. Anno MDCCLI.

Fludd, Robert, alias de Fluctibus (auch Otreb), Clavis Philosophiae et Alchymiae Fluddanae. Francofurti, MDCXXXIII.

——Sophiae cum Moria Certamen. M DC XXIX.

——Tractatus theologico-philosophicus in libros tres distributus, quorum I de Vita, II de Morte, III de Resurrectione. Oppenheim, 1617.

——Schutzschrift für die Aechtheit der Rosenkreuzergesellschaft. Übersetzt von Ada Mah Booz (Dr. A. M. Birkholz). Leipzig, 1782.

Frizius, Ioachimus, Summum Bonum, quod est verum Magiae, Cabalae, Alchymiae verae Subjectum Fratrum Roseae Crucis verorum. (Frankfurt), M.DC.XXIX.

Helmont, Ioan. Baptista van, Ortus Medicinae. Ed. IV. Lugduni, M.DC.LXVII.

Henoch v. Dillmann (mod.).

Hermes v. Fleischer, Reitzenstein (mod.).

I. C. H., Des Hermes Trismegists wahrer alter Naturweg. Herausgegeben von einem ächten Freymäurer I. C. H. Leipzig, 1782.

" Kabbala Denudata seu Doctrina Hebraeorum transcendentalis et metaphysica atque theologica." 2 Bde. Sulzbach, 1677; Frankfurt, 1684. (Herausgeber Knorr von Rosenroth.)

Khunrath, Heinrich, Amphitheatrum Sapientiae Aeternae, solius verae, Christiano-Kabalisticum, Divino-Magicum, Tertriunum, Catholicon. Hanoviae, MDCIX.

——Tractat von den ersten Elementen. (Beygefüget: Unterricht für den Adeptengrad.) Herausgegeben von einem Verehrer der edlen Schmelzund Maurerkunst. Leipzig, 1784.

——Wahrhafter Bericht vom philosophischen Athanor. Leipzig, 1783.

Lacinius, Janus, Praeciosa ac nobilissima artis chymiae collectanea. Norimbergae, M.D.LIIII.

——Pretiosa Margarita novella de Thesauro, ac pretiosissimo Philosophorum Lapide. Venetiis, 1546.

Leade, Jane, Der Baum des Glaubens (beigebunden den
"Offenbarungen, die letzten Zeiten betreffend."
Strassburg, 1807).
——Ein Garten-Brunn gewässert durch die Ströhme der
göttlichen Lustbarkeit, und hervorgrünend in man-
nichfaltigem Unterschiede geistlicher Pflanzen. Am-
sterdam, 1697–1700. 3 Tle. (LGB)
——Offenbahrung der Offenbahrungen. Amsterdam, 1695.
Limitibus, Philoteus de, Allgemeine Abbildung der ganzen
Schöpfung. Aus dem Lateinischen übersezt und mit
Anmerkungen begleitet von J. J. Grienstein. Phila-
delphia, 1792.
——Das Hermetische Triklinium oder drei Gespräche vom
Stein der Weisen. Aus dem Lateinischen übersezt und
mit Anmerkung begleitet von J. J. Grienstein. Phila-
delphia, 1792.
Maier, Michael, Arcana Arcanissima. (ca. 1616.)
——Atalanta Fugiens. Oppenheimii, 1618.
——Cantilenae intellectuales de Phoenice redivivo. Tra-
duites en François . . . par M. L. L. M. Paris,
M.DCC.LVIII.
——Symbola Aureae Mensae duodecim nationum. Francof,
1617.
——Tripus Aureus, hoc est, tres tractatus chymici selecti.
Francofurti, 1678.
Mangetus, Jo. Jacobus, Bibliotheca chemica curiosa. Ge-
nevae, M.DCC.II. 2 Bde.
"Manresa" v. mod.
Mosheim, Johann Lorenz, Versuch einer unpartheiischen
und gründlichen Ketzergeschichte. Helmstaedt, 1746.
Mothe-Guyon, La Vie de Madame J. M. B. de la Mothe-
Guyon, écrite par elle-même. Paris, M.DCC.XC. 3
Bde.
"Musaeum hermeticum reformatum et amplificatum, omnes
sophospagyricae artis discipulos fidelissime erudiens."
Francofurti, 1749.
Mystiker, Deutsche, des XIV. Jahrhunderts, v. Pfeiffer
(mod.).
Nicolai, Friedrich, Versuch über die Beschuldigungen welche
dem Tempelherrenorden gemacht worden, und über
dessen Geheimniss; nebst einem Anhange über das

Entstehen der Freimaurergesellschaft. 2 Tle. Berlin und Stettin, 1782.

Otreb v. Fludd.

Paracelsus (A. Ph. Theophrastus Bombastus von Hohenheim), Werke. 4°-Ausg. Huser, Basel, 1589–90. 10 Bde. Fol.-Ausg. Strassburg, 1603 u. 1616–18. 2 Bde.

Patanjali v. mod.

Pernety, Dom Antoine-Joseph, Dictionnaire mytho-hermétique. Paris, M.DCC.LVIII.

⸻Les Fables Egyptiennes et Grecques dévoilées & réduites au même principe. Paris, M.DCC.LXXXVI. 2 Bde.

Philaletha, Des Hochgelehrten Philalethae und anderer auserlesene Chymische Tractätlein . . . ins Teutsche übersetzet von Johann Langen. Wienn, 1749.

Philaletha, Eugenius (Thomas Vaughan), Lumen de Lumine. Ins Teutsche übersetzet von I. R. S. M. C. Hof, 1750.

⸻Magia Adamica oder das Alterthum der Magie. Amsterdam, 1704.

Pistorius, Ioannes, Artis Cabalisticae, hoc est, reconditae theologiae et philosophiae, scriptorum Tomus I. Basileae, MDXIIIC.

Platon, Werke.

Plotinus v. Kiefer (mod.).

Pordage, John (Johann Pordädsch), Göttliche und Wahre Metaphysica. Franckfurt und Leipzig, M DCC XV. 3 Bde.

⸻Gründlich Philosophisch Sendschreiben, Ein, vom rechten und wahren Steine der Weissheit. Amsterdam, 1698.

⸻Sophia, das ist, Die Holdseelige ewige Jungfrau der Göttlichen Weisheit. Amsterdam, 1699.

⸻Theologia Mystica. Amsterdam, 1698.

⸻Vier Tractätlein des Seeligen Johannes Pordädschens M. D. in Manuschriptis hinterlassen. Amsterdam, 1704.

Reuchlin, Ioannes, De Arte Cabalistica libri tres.— De Verbo Mirifico libri III. Basileae, M.D.LXI.

Riplaeus, Georgius (George Ripley), Chymische Schriften.

Ins Teutsche übersetzet durch Benjamin Roth-Scholtzen. Wienn. 1756.

Rulandus, Martinus, Lexicon Alchemiae. In libera Francofurtensium Repub. M D C XII.

Ruysbroeck, Johann van, v. mod.

Samkara v. Deussen (mod.).

S(chröder), Neue Alchymistische Bibliothek für den Naturkundiger unsers Jahrhunderts ausgesucht. 4 Sammlgn in 2 Bdn. Frankf. u. Leipz. 1772.

Sendivogius, Michael, Chymische Schriften. Nebst einem kurzen Vor-bericht ans Liecht gestellet durch Friedrich Roth-Scholtzen. Wienn, 1750.

"Sieben-Gestirn, Alchimistisch, Das ist, Sieben schöne und auserlesene Tractätlein vom Stein der Weisen." Frankfurt am Main, 1756.

Sperber, Julius, Mysterium magnum. Amsterdam, 1660.

——Tractatus, Ein Feiner, von vielerley wunderbarlichen . . . Dingen. Amsterdam, 1662.

(Starck, Johann August) Anonym, Ueber die alten und neuen Mysterien. Berlin, 1782.

Svātmārām Svāmi v. mod.

Tauler, Joannes, Predig, fast fruchtbar zu eim recht christlichen leben. Basel, M.D.XXII.

Tausendeine Nacht v. Erzählungen (mod.).

Theatrum Chemicum, praecipuos selectorum auctorum tractatus de chemiae et lapidis philosophici antiquitate, veritate . . . continens. Argentorati, M.DC.LIX–M.DC.LXI. 6 Bde.

Valentinus, Basilius, Chymische Schriften. In drey Theile verfasset, samt einer neuen Vorrede . . . von Bened. Nic. Petraeo. Leipzig, 1769.

Villars, Abbe de, Le Comte des Gabalis, ou Entretiens sur les Sciences Secrètes. Londres, M.DCC.XLII.

"Wasserstein der Weisen." Vormahlen durch Lucas Jennis ausgegeben, nunmehro aber wiederum neu aufgelegt . . . von dem F. R. C. Frankfurt und Leipzig, 1760.

Welling, Georgius, Opus mago-cabbalisticum et theosophicum. Darinnen der Ursprung, Natur, . . . des Saltzes, Schwefels und Mercurii in dreyen Theilen beschrieben . . . wird. Homburg vor der Höhe, 1735.

" Zohar " v. de Pauly (mod.).
Zosimos v. Berthelot, Hoefer (mod.).

### MODERN. (From 1800 on.)

Abraham, Dr. Karl, Traum und Mythus. Leipzig und
    Wien, 1909. English Translation.
Adler, Dr. Alfred, Uber den nervösen Charakter. Wies-
    baden, 1912. English Translation.
——Die psychische Behandlung der Trigeminusneuralgie.
    Zentralbl. f. Psych., I., Heft 1/2.
——Der psychische Hermaphroditismus in Leben und in der
    Neurose.— Fortschritte der Medizin, Leipzig, 1910,
    Heft 16.
Basile, Giambattista, Das Märchen aller Märchen oder das
    Pentameron. Neu bearb. von Hans Floerke. 2 Bde.
    München u. Leipzig, 1909.
Bastian, Adolf, Der Mensch in der Geschichte. 3 Bde.
    Leipzig, 1860.
Bauer, Prof. A., Chemie und Alchymie in Osterreich bis
    zum beginnenden XIX. Jahrhundert. Wien, 1883.
Baur, Dr. Ferdinand Christian, Die christliche Gnosis.
    Tübingen, 1835.
Berthelot, Marcellin, Collection des anciens Alchimistes
    Grecs, texte et traduction. Avec la collaboration de
    Ch.-M. Ruello. Paris, 1887 à 1888. 3 vols.
——Introduction à l'Etude de la Chimie des anciens et du
    moyen-âge. Paris, 1889.
——Les Origines de l'Alchimie. Paris, 1885.
——Die Chemie im Altertum und im Mittelalter. Uber-
    tragen von Emma Kalliwoda. Mit Anmerkungen von
    Dr. Franz Strunz. Leipzig und Wien, 1909.
Besetzny, Dr. Emil, Die Sphinx. Freimaur. Taschenbuch.
    Wien, 1873.
Bischoff, Dr. Erich, Die Kabbalah. Leipzig, 1903.
Blau, Dr. Ludwig, Das altjüdische Zauberwesen. Strass-
    burg i. E., 1898.
Boas, Franz, Indianische Sagen von der Nord-Pacifischen
    Küste Amerikas. Berlin, 1895.
Bousset, Dr. Wilhelm, Hauptprobleme der Gnosis. Göttin-
    gen, 1907.

Brabbée, Gustav, Sub Rosa. Vertrauliche Mitteilungen aus dem maurerischen Leben unserer Grossväter. Wien, 1879.

Brandeis, J., Sippurim. 2 Aufl. Prag, 1889.

Brandt, Wilhelm, Mandäische Schriften. Göttingen, 1893.

Brugsch, Heinrich, Religion und Mythologie der alten Ägypter. 2. verm. Ausg. Leipzig, 1891.

Buhle, Johann Gottlieb, Über den Ursprung und die vornehmsten Schicksale der Orden der Rosenkreuzer und Freymaurer Göttingen, 1804.

Comenius-Gesellschaft, Monatshefte der C.-G. für Kultur und Geistesleben. Herausgegeben von Ludwig Keller (Berlin). Jena.

Creuzer, Dr. Friedrich, Symbolik und Mythologie der alten Völker. 2. Aufl. 6 Bde. Leipzig und Darmstadt, 1819ff.

Deussen, Dr. Paul, Allgemeine Geschichte der Philosophie. Leipzig.

——Sechzig Upanishad's des Veda. 2. Aufl. Leipzig, 1905.

——Das System des Vedânta. Nach den Brahma-Sûtra's des Bâdarâyana und dem Kommentare des Çañkara über dieselben. 2. Aufl. Leipzig, 1906.

——Die Sûtra's des Vedânta . . . nebst dem vollständigen Commentare des Çankara. Leipzig, 1887.

Dieterich, Albrecht, Mutter Erde. Leipzig, 1905.

Dillmann, Dr. A., Das Buch Henoch. Leipzig, 1853.

Dulaure, Jacques-Antoine, Die Zeugung in Glauben, Sitten und Bräuchen der Völker. Verdeutscht und ergänzt von Friedrich S. Krauss und Karl Reiskel. Leipzig, 1909.

Ehrenreich, Paul, Die allgemeine Mythologie und ihre ethnologischen Grundlagen. Leipzig, 1910.

Ennemoser, Dr. Joseph, Geschichte der Magie. Leipzig, 1844.

Erman, Adolf, Die Agyptische Religion. Berlin, 1905.

"Erzählungen aus den Tausend und ein Nächten," Die. Ausgabe Felix Paul Greve. Leipzig, MDCCCCVII–MDCCCCVIII. 12 Bde.

Ferenczi, Dr. Sándor, Introjektion und Übertragung. Jb. ps. F. I. 2. English Translation.

——Symbolische Darstellung des Lust-und Realitätsprinzips im Oedipus-Mythos. Imago, I. 3.

Ferrero, Guillaume, Les Lois psychologiques du Symbolisme. Paris, 1895.

Fichte, Johann Gottlieb, Die Anweisung zum seeligen Leben. Berlin, 1806.

Findel, J. G., Geschichte der Freimaurerei. 3. Aufl. Leipzig, 1870.

Fischer, Robert, Erläuterung der Katechismen der Joh.-Freimaurerei. I–IV.

Fleischer, Prof. Dr. H. L., Hermes Trismegistus an die menschliche Seele. Leipzig, 1870.

Flournoy, Th., Des Indes à la Planète Mars. 4. éd. Paris.

Franck, Ad., La Kabbale. Paris, 1892.

Freimark, Hans, Moderne Theosophen und ihre Theosophie. Leipz., 1912.

——Okkultismus und Sexualität. Leipzig, 1909.

——Okkultistische Bewegung, Die. Leipzig, 1912.

Freud, Prof. Dr. Sigmund, Drei Abhandlungen zur Sexualtheorie. Leipzig und Wien, 1905. English Translation.

——Dynamik der Übertragung, Zur. Zentralbl. f. Psych. II. 4.

——Formulierungen über die zwei Prinzipien des psychischen Geschehens. Jb. ps. F. III. 1.

——Psychopathologie des Alltagslebens, Zur. 3. Aufl. Berlin, 1910. English Translation.

——Sammlung kleiner Schriften zur Neurosenlehre. 2 Bde. Leipzig und Wien, 1906, 1909.

——Traumdeutung, Die 3. Aufl. Leipzig und Wien, 1911. English Translation.

——Wahn, Der, und die Träume in W. Jensens "Gradiva." Leipzig und Wien, 1907. English Translation.

Frobenius, Leo, Das Zeitalter des Sonnengottes. I. Bd. Berlin, 1904.

Furtmüller, Dr. Karl, Psychoanalyse und Ethik. München, 1912.

Garbe, Richard, Die Sāmkhya-Philosophie. Leipzig, 1894.

——Sāmkhya und Yoga. Grundriss der Indo-Arischen

Philologie, herausgegeben von Georg Bühler, III. 4.
Strassburg, 1896.
Gneiting, J. M. (J. Krebs), Maurerische Mitteilungen. 4
Bändchen. Stuttgart, 1831–1833.
Goldziher, Ignaz, Der Mythus bei den Hebräern. Leipzig,
1876.
Görres, J., Die christliche Mystik. 4 Bde. Regensburg,
1836–1842.
Grimm, Brüder, Deutsche Sagen (D S).
——Kinder- und Hausmärchen (KHM).
——Jacob, Deutsche Mythologie. 4. Ausg., von Elard
Hugo Meyer. 3 Bde. Gütersloh, 1876–1878.
Haltrich, Josef, Deutsche Volksmärchen aus dem Sachsen-
lande in Siebenbürgen. 4. Aufl. Wien, Hermann-
stadt, 1885. (Siebenbg.-deutsche Volksbücher, Bd. II.)
Hartmann, Dr. Franz, Secret Symbols of the Rosicrucians.
Boston, 1888. (Eine mangelhafte englische Wieder-
gabe des deutschen Werkes: " Geheime Figuren der
Rosenkreuzer.")
Havelock Ellis, Die Welt der Träume. Deutsche Ausgabe
von Dr. Hans Kurella. Würzburg, 1911.
(Hitchcock Ethan Allen) Anonym, The Story of the Red
Book of Appin. (Als Vervollständigung dazu:) Ap-
pendix to the Story . . . (Beides:) New York, 1863.
——Remarks upon Alchemy and the Alchemists. Boston,
1857. (HA)
——Swedenborg, a Hermetic Philosopher. Being a sequel
to " Remarks on Alchemy and the Alchemists." New
York, 1858.
——Das rote Buch von Appin. Übertr. von Sir Galahad.
Leipzig, 1910.
Hoefer, Ferdinand, Histoire de la Chimie. 2 vols. Paris,
1842–1844.
Höhler, Wilhelm, Hermetische Philosophie und Freimaure-
rei. Ludwigshafen am Rhein, 1905.
Hossbach, Wilhelm, Johann Valentin Andreä und sein Zei-
talter. Berlin, 1819.
Horst, Georg Conrad, Dämonomagie, oder Geschichte des
Glaubens an Zauberei. 2 Tle. Frankfurt a. M.,
1818.
——Deuteroskopie. 2 Bdchen. Frankfurt a. M., 1830.

——Zauber-Bibliothek. 6 Bde. Mainz, 1821 ff.

Horten, M., Mystische Texte aus dem Islam. Drei Gedichte des Arabi 1240. Kleine Texte f. Vorlesgn. u. Übgn., herausg. von Hans Lietzmann, Nr. 105. Bonn, 1912.

"Imago," Zeitschrift für Anwendung der Psychoanalyse auf mund Freud. Wien, 1912f. See Psychoanalytic Review.

Inman, Thomas M. D., Ancient Pagan and Modern Christian Symbolism. 4. ed. New York, 1884.

Janet, Pierre, L'Automatisme Psychologique. Paris, 1889.

——Les Névroses. Paris, 1910.

"Jahrbuch für psychoanalytische und psychopathologische Forschungen." Herausgegeben von Prof. Dr. E. Bleuler und Prof. Dr. S. Freud. Leipzig und Wien, 1909ff. (Jb. ps. F.) See Psychoanalytic Review.

Jeremias, Alfred, Alte Testament, Das, im Lichte des Alten Orients. 2. Aufl. Leipzig, 1906.

——Babylonisch-assyrischen Vorstellungen vom Leben nach dem Tode, Die. Leipzig, 1887.

Jodl, Friedrich, Geschichte der Ethik. 2 Bde. 2. Aufl. Stuttgart und Berlin, 1906, 1912.

Jung, Dr. C. G., Bedeutung des Vaters für das Schicksal des Einzelnen, Die. Jb. ps. F. I. 2. English Translation.

——Konflikte der kindlichen Seele, Über. Jb. ps. F. II. 1. English Translation.

——Psychologie und Pathologie sogenannter occulter Phänomene, Zur. Leipzig, 1902. English Translation.

——Wandlungen und Symbole der Libido. Beiträge zur Entwicklungsgeschichte des Denkens. Jb. ps. F. III.–IV. English Translation.

Katsch, Dr. Ferdinand, Die Entstehung und der wahre Endzweck der Freimaurerei. Berlin, 1897.

Keller, Dr. Ludwig, Anfänge der Renaissance, Die, und die Kultgesellschaften des Humanismus im XIII. und XIV. Jahrhundert. Leipzig und Jena, 1903.

——Bibel, Winkelmass und Zirkel. Jena, 1910.

——Geistigen Grundlagen der Freimaurerei, Die, und das öffentliche Leben. Jena, 1911.

Keller, Dr. Ludwig, Geschichte der Bauhütten und der Hüttengeheimnisse, Zur. Leipzig und Jena, 1898.

——Grossloge Indissolubilis, Die, und andere Grosslogen-
systeme des 16., 17. und 18. Jahrhunderts. Jena, 1908.
——Heiligen Zahlen, Die, und die Symbolik der Katakom-
ben. Leipzig und Jena, 1906.
——Idee der Humanität, Die, und die Comenius-Gesell-
schaft. Jena, 1908.
——Italienischen Akademien des 18. Jahrhunderts, Die, und
die Anfänge des Maurerbundes in den romanischen und
den nordischen Ländern. Leipzig und Jena, 1905.
——Latomien und Loggien in alter Zeit. Leipzig und
Jena, 1906.
——Römische Akademie, Die, und die altchristlichen Kata-
komben im Zeitalter der Renaissance. Leipzig und
Jena, 1899.
——Sozietäten des Humanismus und die Sprachgesellschaf-
ten, Die. Jena, 1909.
——Tempelherren, Die, und die Freimaurer. Leipzig und
Jena, 1905.
Kernning, J. (J. Krebs), Schlüssel zur Geisterwelt oder die
Kunst des Lebens. Neue Auflage. Stuttgart, 1855.
Kiefer, Otto: Plotin, Enneaden. In Auswahl. 2 Bde.
Jena u. Leipz., 1905.
Kiesewetter, Carl, Die Geheimwissenschaften. Leipzig,
1895.
Kleinpaul, Dr. Rudolf, Das Leben der Sprache und ihre
Weltstellung. 3 Bde. Leipzig, 1893.
——Die Lebendigen und die Toten in Volksglauben, Reli-
gion und Sage. Leipzig, 1898.
" Kloster " v. Scheible.
Knight, Richard Payne, The symbolical Language of an-
cient Art. New ed. New York, 1892.
——Le Culte de Priape et ses Rapports avec la Théologie
mystique des Anciens . . . Traduits de l'anglais par
E. W. Bruxelles, 1883.
Kopp, Hermann, Die Alchemie in älterer und neuerer Zeit.
2 Tle. Heidelberg, 1886.
——Geschichte der Chemie. 4 Tle. Braunschweig, 1843–
1847.
————Beiträge zur. 3 Stücke. Braunschweig, 1869
und 1875.
Krause, Karl Christian Friedrich, Die drei ältesten Kuns-

turkunden der Freimaurerbrüderschaft. 2 (4) Bde. Dresden, 1820–1821.

Krauss, Dr. Friedrich S., ʼΑΝΘΡΩΠΟΦΥΤΕΙΑ. Jahrbücher für folkloristische Erhebungen und Forschungen zur Entwicklungsgeschichte der geschlechtlichen Moral. Bisher 9 Bände. Leipzig.

Kuhn, Adalbert, Die Herabkunft des Feuers und des Göttertranks. Berlin, 1859.

Laistner, Ludwig, Das Rätsel der Sphinx. 2 Bde. Berlin, 1889.

Lehmann, Dr. Alfred, Aberglaube und Zauberei. Deutsche Ausg. v. Dr. Petersen. Stuttgart, 1898.

Lenning, C. (Hesse) und Fr(iedrich) M(ossdorf), Encyklopädie der Freimaurerei. III. Auflage = Allgemeines Handbuch der Freimaurerei. 2 Bde. Leipzig, 1900–1901.

Lessmann, Heinrich, Aufgaben und Ziele der vergleichenden Mythenforschung. Leipzig, 1908.

Lévi, Eliphas, Histoire de la Magie. Nouvelle éd. Paris, 1892.

Lipps, G. F., Mythenbildung und Erkenntnis. Leipzig und Berlin, 1907.

Lisco, D. Friedrich Gustav, Die Heilslehre der Theologia deutsch. Stuttgart, 1857.

Lorenz, Dr. Emil, Das Titanen-Motiv in der allgemeinen Mythologie. Imago, II. 1.

Mackenzie, Kenneth R. H., The Royal Masonic Cyclopaedia. London, 1877.

Maeder, Dr. A., Über die Funktion des Traumes. Jb. ps. F. IV. 2. English Translation.

Maimon, Salomon, Lebensgeschichte. Herausg. von Dr. Jakob Fromer. München, 1911.

Mannhardt, Wilhelm, Germanische Mythen. 2 Bde. Berlin, 1858.

" Manresa " oder die geistlichen Übungen des heiligen Ignatius. Nach dem Französischen, von Franz A. Schmid. 6. Aufl. Regensburg, 1903.

Marchand, R. F., Über die Alchemie. Halle, 1847.

Müller, Dr. Friedrich, Siebenbürgische Sagen. 2. Aufl. Wien, Hermannstadt, 1885. (Siebenbg.-deutsche Volksbücher, Bd. I.)

———Karl Otfried, Prolegomena einer wissenschaftlichen Mythologie. Göttingen, 1825.
Nettelbladt, C. C. F. W. Freih. von, Geschichte der freimaurerischen Systeme. Berlin, 1879.
Nietzsche, Werke. Leipzig.
Nork, F., Mythologie der Volkssagen und Volksmärchen. Stuttgart, 1848. (Kloster, Bd. IX.)
———Sitten und Gebräuche der Deutschen und ihrer Nachbarvölker, Die. Stuttgart, 1849. (Kloster, Bd. XII.)
Patanjali, Yoga-Sutra. Translation with introduction, appendix, and notes by Manilal Nabubhai Dvivedi. Bombay, 1890.
Papus (Dr. Gérard Encausse), Traité élémentaire de Magie Pratique. Paris, 1906.
Pauly, Jean de, Sepher ha-Zohar (Le Livre de la Splendeur). Publié par les soins de Emile Lafuma-Giraud. Paris, 1906ff. 6 (7) Bde.
Pfeiffer, Franz, Deutsche Mystiker des vierzehnten Jahrhunderts. 2 Bde. Leipzig, 1845, 1857.
———Theologia deutsch. 2. Aufl. Stuttgart, 1855.
Pfister, Dr. Oskar, Die Frömmigkeit des Grafen Ludwig von Zinzendorf. Leipzig und Wien, 1910.
Poisson, Albert, Théories et Symboles des Alchimistes. Paris, 1891.
Prasád, Ráma, Nature's Finer Forces. London, 1890.
Rank, Dr. Otto, Inzest-Motiv in Dichtung und Sage, Das. Leipzig und Wien, 1912.
———Künstler, Der. Wien, 1907.
———Mythus von der Geburt des Helden, Der. Leipzig und Wien, 1909. English Translation.
———Lohengrinsage, Die. Leipzig und Wien, 1911.
———Symbolschichtung im Wecktraum, Die, und ihre Wiederkehr im mythischen Denken. Jb. ps. F. IV. 1.
———Völkerpsychologische Parallelen zu den infantilen Sexualtheorien. Zentralbl. f. Psych. II. 7–8.
Reitzenstein, R., Hellenistischen Mysterienreligionen, Die. Leipzig, 1910.
———Poimandres. Leipzig, 1904.
Riklin, Dr. Franz, Wunscherfüllung und Symbolik im Märchen. Leipzig und Wien, 1908. English Translation.

——(Vorträge). Zentralbl. f. Psychoanalyse III. 2., pag. 103ff., 113ff.

Roscher, W. H., Ausführliches Lexikon der Griechischen und Römischen Mythologie. Leipzig, 1884ff.

Ruysbroeck, Johann van, Drei Schriften. Aus dem Vlämischen von Franz A. Lambert. Leipzig.

Scheible, J., Das Kloster. Meist aus der ältern deutschen Volks-, Wunder-, Curiositäten-, und vorzugsweise komischen Literatur. 12 Bde. Stuttgart, 1845–1849.

Scherner, R. A., Das Leben des Traums. Berlin, 1861.

Schiffmann, G. A., Die Freimaurerei in Frankreich in der ersten Hälfte des XVIII. Jahrhunderts. Leipzig, 1881.

Schlegel, Aug. Guil. a, Bhagavad-Gita id est Θεσπεσιον Μελος . . . Editio altera, cura Christiani Lasseni. Bonnae, MDCCCXLVI.

Schmieder, Karl Christoph, Geschichte der Alchemie. Halle, 1832.

Schmidt, Richard, Fakire und Fakirtum im alten und modernen Indien. Berlin, 1908.

Schneider, Hermann, Kultur und Denken der alten Agypter. Leipzig, 1907.

Schopenhauer, Werke.

Schroeder, Leopold von, Bhagavad-Gita. Des Erhabenen Sang. Jena, 1912.

Schubert, Dr. G. H. von, Die Symbolik des Traumes. 3. Aufl. Leipzig, 1840.

Silberer, Herbert, Charakteristik des lekanomantischen Schauens, Zur. Zentralbl. f. Psych. III. 2–3.

——Kategorien der Symbolik, Von den. Zentralbl. f. Psych. II. 4.

——Lekanomantische Versuche. Zentralbl. f. Psych. II. 7–10.

——Mantik und Psychanalyse. Zentralbl. f. Psych. II. 2.

——Phantasie und Mythos. Vornehmlich vom Gesichtspunkte der funktionalen Kategorie aus betrachtet. Jb. ps. F. II. 2.

——Prinzipielle Anregung, Eine. Jb. ps. F. IV. 2.

——Spermatozoenträume, Zur Frage der. Jb. ps. F. IV. 2.

——Symbolbildung. Jb. ps. F. III. 2., IV. 2.

——Symbolik des Erwachens und Schwellensymbolik über-
haupt. Jb. ps. F. III. 2.
Soldan, Geschichte der Hexenprozesse. Neu bearbeitet von
Dr. Heinrich Heppe. 2 Bde. Stuttgart, 1880.
Spielrein, Dr. Sabina, Über den psychologischen Inhalt eines
Falles von Schizophrenie (Dementia praecox). Jb. ps.
F. III. 1.
——Die Destruktion als Ursache des Werdens. Jb. ps. F.
IV. 1.
Starcke, Dr. C. N., Freimaurerei als Lebenskunst. Berlin,
1911.
Staudenmaier, Dr. Ludwig, Die Magie als experimentelle
Naturwissenschaft. Leipzig, 1912.
Stekel, Dr. Wilhelm, Nervöse Angstzustände und ihre Be-
handlung. 2. Aufl. Berlin und Wien, 1912.
——Sprache des Traumes, Die. Wiesbaden, 1911.
——Traumdeutung, Fortschritte in der. Zentralbl. f.
Psych. III.
——Träume der Dichter, Die. Wiesbaden, 1912.
Strunz, Dr. Franz, Naturbetrachtung und Naturerkenntnis
im Altertum. Hamburg und Leipzig, 1904.
——Theophrastus Paracelsus. Jena, 1903.
Stucken, Eduard, Astralmythen. Religionsgeschichtliche
Untersuchungen. Leipzig, 1907. (S A M)
Svātmārām Svāmi, Hatha-Yoga Pradīpikā. Translated by
Shrinivās Iyāngār. Published by Tookaram Tatya.
Bombay, 1885.
(Wirth, Oswald) Anonym, Le Livre de l'Apprenti. Nouv.
éd. Publié par la L .·. Travail & Vrais Amis Fidèles.
Paris, 1898.
——Le Symbolisme Hermétique dans ses Rapports avec la
Franc-Maçonnerie. Paris, 1909. (W S H)
Wundt, Wilhelm, Völkerpsychologie (Mythus und Reli-
gion). 3 Tle. Leipzig, 1910 (I. Teil, 2. Aufl.),
1906, 1909. English Translation.
Zeitschrift für ärztliche Psychoanalyse, Internationale.
Herausg. von Prof. Dr. Sigmund Freud. Leipzig und
Wien, 1913. See Psychoanalytic Review.
Zentralblatt für Psychoanalyse. Bd. I–II, herausg. von.
Prof. Dr. Freud. Bd. IIIf. von Dr. Wilhelm Stekel.

Wiesbaden, 1910ff. (Zb. f. Ps.) See Psychoanalytic Review.

**Note.** The works of Freud, Adler, Jung, Rank, and Ricklin, are to be found in English Translations. See Psychoanalytic Review, N. Y., Nervous and Mental Disease Monograph Series and lists, Moffat, Yard & Co., N. Y.

# INDEX

A CATALOGUE OF SELECTED DOVER BOOKS
IN ALL FIELDS OF INTEREST

# A CATALOGUE OF SELECTED DOVER
# BOOKS IN ALL FIELDS OF INTEREST

RACKHAM'S COLOR ILLUSTRATIONS FOR WAGNER'S RING. Rackham's finest mature work—all 64 full-color watercolors in a faithful and lush interpretation of the *Ring*. Full-sized plates on coated stock of the paintings used by opera companies for authentic staging of Wagner. Captions aid in following complete Ring cycle. Introduction. 64 illustrations plus vignettes. 72pp. 8⅝ x 11¼. 23779-6 Pa. $6.00

CONTEMPORARY POLISH POSTERS IN FULL COLOR, edited by Joseph Czestochowski. 46 full-color examples of brilliant school of Polish graphic design, selected from world's first museum (near Warsaw) dedicated to poster art. Posters on circuses, films, plays, concerts all show cosmopolitan influences, free imagination. Introduction. 48pp. 9⅜ x 12¼. 23780-X Pa. $6.00

GRAPHIC WORKS OF EDVARD MUNCH, Edvard Munch. 90 haunting, evocative prints by first major Expressionist artist and one of the greatest graphic artists of his time: *The Scream, Anxiety, Death Chamber, The Kiss, Madonna,* etc. Introduction by Alfred Werner. 90pp. 9 x 12. 23765-6 Pa. $5.00

THE GOLDEN AGE OF THE POSTER, Hayward and Blanche Cirker. 70 extraordinary posters in full colors, from Maitres de l'Affiche, Mucha, Lautrec, Bradley, Cheret, Beardsley, many others. Total of 78pp. 9⅜ x 12¼. 22753-7 Pa. $5.95

THE NOTEBOOKS OF LEONARDO DA VINCI, edited by J. P. Richter. Extracts from manuscripts reveal great genius; on painting, sculpture, anatomy, sciences, geography, etc. Both Italian and English. 186 ms. pages reproduced, plus 500 additional drawings, including studies for *Last Supper,* Sforza monument, etc. 860pp. 7⅞ x 10¾. (Available in U.S. only) 22572-0, 22573-9 Pa., Two-vol. set $15.90

THE CODEX NUTTALL, as first edited by Zelia Nuttall. Only inexpensive edition, in full color, of a pre-Columbian Mexican (Mixtec) book. 88 color plates show kings, gods, heroes, temples, sacrifices. New explanatory, historical introduction by Arthur G. Miller. 96pp. 11⅜ x 8½. (Available in U.S. only) 23168-2 Pa. $7.95

UNE SEMAINE DE BONTÉ, A SURREALISTIC NOVEL IN COLLAGE, Max Ernst. Masterpiece created out of 19th-century periodical illustrations, explores worlds of terror and surprise. Some consider this Ernst's greatest work. 208pp. 8⅛ x 11. 23252-2 Pa. $6.00

THE DEPRESSION YEARS AS PHOTOGRAPHED BY ARTHUR ROTH-STEIN, Arthur Rothstein. First collection devoted entirely to the work of outstanding 1930s photographer: famous dust storm photo, ragged children, unemployed, etc. 120 photographs. Captions. 119pp. 9¼ x 10¾.
23590-4 Pa. $5.00

CAMERA WORK: A PICTORIAL GUIDE, Alfred Stieglitz. All 559 illustrations and plates from the most important periodical in the history of art photography, Camera Work (1903-17). Presented four to a page, reduced in size but still clear, in strict chronological order, with complete captions. Three indexes. Glossary. Bibliography. 176pp. 8⅜ x 11¼.
23591-2 Pa. $6.95

ALVIN LANGDON COBURN, PHOTOGRAPHER, Alvin L. Coburn. Revealing autobiography by one of greatest photographers of 20th century gives insider's version of Photo-Secession, plus comments on his own work. 77 photographs by Coburn. Edited by Helmut and Alison Gernsheim. 160pp. 8⅛ x 11.
23685-4 Pa. $6.00

NEW YORK IN THE FORTIES, Andreas Feininger. 162 brilliant photographs by the well-known photographer, formerly with Life magazine, show commuters, shoppers, Times Square at night, Harlem nightclub, Lower East Side, etc. Introduction and full captions by John von Hartz. 181pp. 9¼ x 10¾.
23585-8 Pa. $6.95

GREAT NEWS PHOTOS AND THE STORIES BEHIND THEM, John Faber. Dramatic volume of 140 great news photos, 1855 through 1976, and revealing stories behind them, with both historical and technical information. Hindenburg disaster, shooting of Oswald, nomination of Jimmy Carter, etc. 160pp. 8¼ x 11.
23667-6 Pa. $5.00

THE ART OF THE CINEMATOGRAPHER, Leonard Maltin. Survey of American cinematography history and anecdotal interviews with 5 masters—Arthur Miller, Hal Mohr, Hal Rosson, Lucien Ballard, and Conrad Hall. Very large selection of behind-the-scenes production photos. 105 photographs. Filmographies. Index. Originally Behind the Camera. 144pp. 8¼ x 11.
23686-2 Pa. $5.00

DESIGNS FOR THE THREE-CORNERED HAT (LE TRICORNE), Pablo Picasso. 32 fabulously rare drawings—including 31 color illustrations of costumes and accessories—for 1919 production of famous ballet. Edited by Parmenia Migel, who has written new introduction. 48pp. 9⅜ x 12¼. (Available in U.S. only)
23709-5 Pa. $5.00

NOTES OF A FILM DIRECTOR, Sergei Eisenstein. Greatest Russian filmmaker explains montage, making of Alexander Nevsky, aesthetics; comments on self, associates, great rivals (Chaplin), similar material. 78 illustrations. 240pp. 5⅜ x 8½.
22392-2 Pa. $4.50

AMERICAN BIRD ENGRAVINGS, Alexander Wilson et al. All 76 plates. from Wilson's *American Ornithology* (1808-14), most important ornithological work before Audubon, plus 27 plates from the supplement (1825-33) by Charles Bonaparte. Over 250 birds portrayed. 8 plates also reproduced in full color. 111pp. 9⅜ x 12½. 23195-X Pa. $6.00

CRUICKSHANK'S PHOTOGRAPHS OF BIRDS OF AMERICA, Allan D. Cruickshank. Great ornithologist, photographer presents 177 closeups, groupings, panoramas, flightings, etc., of about 150 different birds. Expanded *Wings in the Wilderness*. Introduction by Helen G. Cruickshank. 191pp. 8¼ x 11. 23497-5 Pa. $6.00

AMERICAN WILDLIFE AND PLANTS, A. C. Martin, et al. Describes food habits of more than 1000 species of mammals, birds, fish. Special treatment of important food plants. Over 300 illustrations. 500pp. 5⅜ x 8½. 20793-5 Pa. $4.95

THE PEOPLE CALLED SHAKERS, Edward D. Andrews. Lifetime of research, definitive study of Shakers: origins, beliefs, practices, dances, social organization, furniture and crafts, impact on 19th-century USA, present heritage. Indispensable to student of American history, collector. 33 illustrations. 351pp. 5⅜ x 8½. 21081-2 Pa. $4.50

OLD NEW YORK IN EARLY PHOTOGRAPHS, Mary Black. New York City as it was in 1853-1901, through 196 wonderful photographs from N.-Y. Historical Society. Great Blizzard, Lincoln's funeral procession, great buildings. 228pp. 9 x 12. 22907-6 Pa. $8.95

MR. LINCOLN'S CAMERA MAN: MATHEW BRADY, Roy Meredith. Over 300 Brady photos reproduced directly from original negatives, photos. Jackson, Webster, Grant, Lee, Carnegie, Barnum; Lincoln; Battle Smoke, Death of Rebel Sniper, Atlanta Just After Capture. Lively commentary. 368pp. 8⅜ x 11¼. 23021-X Pa. $8.95

TRAVELS OF WILLIAM BARTRAM, William Bartram. From 1773-8, Bartram explored Northern Florida, Georgia, Carolinas, and reported on wild life, plants, Indians, early settlers. Basic account for period, entertaining reading. Edited by Mark Van Doren. 13 illustrations. 141pp. 5⅜ x 8½. 20013-2 Pa. $5.00

THE GENTLEMAN AND CABINET MAKER'S DIRECTOR, Thomas Chippendale. Full reprint, 1762 style book, most influential of all time; chairs, tables, sofas, mirrors, cabinets, etc. 200 plates, plus 24 photographs of surviving pieces. 249pp. 9⅞ x 12¾. 21601-2 Pa. $7.95

AMERICAN CARRIAGES, SLEIGHS, SULKIES AND CARTS, edited by Don H. Berkebile. 168 Victorian illustrations from catalogues, trade journals, fully captioned. Useful for artists. Author is Assoc. Curator, Div. of Transportation of Smithsonian Institution. 168pp. 8½ x 9½. 23328-6 Pa. $5.00

PRINCIPLES OF ORCHESTRATION, Nikolay Rimsky-Korsakov. Great classical orchestrator provides fundamentals of tonal resonance, progression of parts, voice and orchestra, tutti effects, much else in major document. 330pp. of musical excerpts. 489pp. 6½ x 9¼.　　21266-1 Pa. $7.50

TRISTAN UND ISOLDE, Richard Wagner. Full orchestral score with complete instrumentation. Do not confuse with piano reduction. Commentary by Felix Mottl, great Wagnerian conductor and scholar. Study score. 655pp. 8⅛ x 11.　　22915-7 Pa. $13.95

REQUIEM IN FULL SCORE, Giuseppe Verdi. Immensely popular with choral groups and music lovers. Republication of edition published by C. F. Peters, Leipzig, n. d. German frontmaker in English translation. Glossary. Text in Latin. Study score. 204pp. 9⅜ x 12¼.
23682-X Pa. $6.00

COMPLETE CHAMBER MUSIC FOR STRINGS, Felix Mendelssohn. All of Mendelssohn's chamber music: Octet, 2 Quintets, 6 Quartets, and Four Pieces for String Quartet. (Nothing with piano is included). Complete works edition (1874-7). Study score. 283 pp. 9⅜ x 12¼.
23679-X Pa. $7.50

POPULAR SONGS OF NINETEENTH-CENTURY AMERICA, edited by Richard Jackson. 64 most important songs: "Old Oaken Bucket," "Arkansas Traveler," "Yellow Rose of Texas," etc. Authentic original sheet music, full introduction and commentaries. 290pp. 9 x 12.　　23270-0 Pa. $7.95

COLLECTED PIANO WORKS, Scott Joplin. Edited by Vera Brodsky Lawrence. Practically all of Joplin's piano works—rags, two-steps, marches, waltzes, etc., 51 works in all. Extensive introduction by Rudi Blesh. Total of 345pp. 9 x 12.　　23106-2 Pa. $14.95

BASIC PRINCIPLES OF CLASSICAL BALLET, Agrippina Vaganova. Great Russian theoretician, teacher explains methods for teaching classical ballet; incorporates best from French, Italian, Russian schools. 118 illustrations. 175pp. 5⅜ x 8½.　　22036-2 Pa. $2.50

CHINESE CHARACTERS, L. Wieger. Rich analysis of 2300 characters according to traditional systems into primitives. Historical-semantic analysis to phonetics (Classical Mandarin) and radicals. 820pp. 6⅛ x 9¼.
21321-8 Pa. $10.00

EGYPTIAN LANGUAGE: EASY LESSONS IN EGYPTIAN HIERO-GLYPHICS, E. A. Wallis Budge. Foremost Egyptologist offers Egyptian grammar, explanation of hieroglyphics, many reading texts, dictionary of symbols. 246pp. 5 x 7½. (Available in U.S. only)
21394-3 Clothbd. $7.50

AN ETYMOLOGICAL DICTIONARY OF MODERN ENGLISH, Ernest Weekley. Richest, fullest work, by foremost British lexicographer. Detailed word histories. Inexhaustible. Do not confuse this with Concise Etymological Dictionary, which is abridged. Total of 856pp. 6½ x 9¼.
21873-2, 21874-0 Pa., Two-vol. set $12.00

THE COMPLETE WOODCUTS OF ALBRECHT DURER, edited by Dr. W. Kurth. 346 in all: "Old Testament," "St. Jerome," "Passion," "Life of Virgin," Apocalypse," many others. Introduction by Campbell Dodgson. 285pp. 8½ x 12¼.　21097-9 Pa. $7.50

DRAWINGS OF ALBRECHT DURER, edited by Heinrich Wolfflin. 81 plates show development from youth to full style. Many favorites; many new. Introduction by Alfred Werner. 96pp. 8⅛ x 11.　22352-3 Pa. $5.00

THE HUMAN FIGURE, Albrecht Dürer. Experiments in various techniques—stereometric, progressive proportional, and others. Also life studies that rank among finest ever done. Complete reprinting of Dresden Sketchbook. 170 plates. 355pp. 8⅜ x 11¼.　21042-1 Pa. $7.95

OF THE JUST SHAPING OF LETTERS, Albrecht Dürer. Renaissance artist explains design of Roman majuscules by geometry, also Gothic lower and capitals. Grolier Club edition. 43pp. 7⅞ x 10¾　21306-4 Pa. $3.00

TEN BOOKS ON ARCHITECTURE, Vitruvius. The most important book ever written on architecture. Early Roman aesthetics, technology, classical orders, site selection, all other aspects. Stands behind everything since. Morgan translation. 331pp. 5⅜ x 8½.　20645-9 Pa. $4.50

THE FOUR BOOKS OF ARCHITECTURE, Andrea Palladio. 16th-century classic responsible for Palladian movement and style. Covers classical architectural remains, Renaissance revivals, classical orders, etc. 1738 Ware English edition. Introduction by A. Placzek. 216 plates. 110pp. of text. 9½ x 12¾.　21308-0 Pa. $10.00

HORIZONS, Norman Bel Geddes. Great industrialist stage designer, "father of streamlining," on application of aesthetics to transportation, amusement, architecture, etc. 1932 prophetic account; function, theory, specific projects. 222 illustrations. 312pp. 7⅞ x 10¾.　23514-9 Pa. $6.95

FRANK LLOYD WRIGHT'S FALLINGWATER, Donald Hoffmann. Full, illustrated story of conception and building of Wright's masterwork at Bear Run, Pa. 100 photographs of site, construction, and details of completed structure. 112pp. 9¼ x 10.　23671-4 Pa. $5.50

THE ELEMENTS OF DRAWING, John Ruskin. Timeless classic by great Viltorian; starts with basic ideas, works through more difficult. Many practical exercises. 48 illustrations. Introduction by Lawrence Campbell. 228pp. 5⅜ x 8½.　22730-8 Pa. $3.75

GIST OF ART, John Sloan. Greatest modern American teacher, Art Students League, offers innumerable hints, instructions, guided comments to help you in painting. Not a formal course. 46 illustrations. Introduction by Helen Sloan. 200pp. 5⅜ x 8½.　23435-5 Pa. $4.00

THE PHILOSOPHY OF HISTORY, Georg W. Hegel. Great classic of Western thought develops concept that history is not chance but a rational process, the evolution of freedom. 457pp. 5⅜ x 8½. 20112-0 Pa. $4.50

LANGUAGE, TRUTH AND LOGIC, Alfred J. Ayer. Famous, clear introduction to Vienna, Cambridge schools of Logical Positivism. Role of philosophy, elimination of metaphysics, nature of analysis, etc. 160pp. 5⅜ x 8½. (Available in U.S. only) 20010-8 Pa. $2.00

A PREFACE TO LOGIC, Morris R. Cohen. Great City College teacher in renowned, easily followed exposition of formal logic, probability, values, logic and world order and similar topics; no previous background needed. 209pp. 5⅜ x 8½. 23517-3 Pa. $3.50

REASON AND NATURE, Morris R. Cohen. Brilliant analysis of reason and its multitudinous ramifications by charismatic teacher. Interdisciplinary, synthesizing work widely praised when it first appeared in 1931. Second (1953) edition. Indexes. 496pp. 5⅜ x 8½. 23633-1 Pa. $6.50

AN ESSAY CONCERNING HUMAN UNDERSTANDING, John Locke. The only complete edition of enormously important classic, with authoritative editorial material by A. C. Fraser. Total of 1176pp. 5⅜ x 8½. 20530-4, 20531-2 Pa., Two-vol. set $16.00

HANDBOOK OF MATHEMATICAL FUNCTIONS WITH FORMULAS, GRAPHS, AND MATHEMATICAL TABLES, edited by Milton Abramowitz and Irene A. Stegun. Vast compendium: 29 sets of tables, some to as high as 20 places. 1,046pp. 8 x 10½. 61272-4 Pa. $14.95

MATHEMATICS FOR THE PHYSICAL SCIENCES, Herbert S. Wilf. Highly acclaimed work offers clear presentations of vector spaces and matrices, orthogonal functions, roots of polynomial equations, conformal mapping, calculus of variations, etc. Knowledge of theory of functions of real and complex variables is assumed. Exercises and solutions. Index. 284pp. 5⅝ x 8¼. 63635-6 Pa. $5.00

THE PRINCIPLE OF RELATIVITY, Albert Einstein et al. Eleven most important original papers on special and general theories. Seven by Einstein, two by Lorentz, one each by Minkowski and Weyl. All translated, unabridged. 216pp. 5⅜ x 8½. 60081-5 Pa. $3.50

THERMODYNAMICS, Enrico Fermi. A classic of modern science. Clear, organized treatment of systems, first and second laws, entropy, thermodynamic potentials, gaseous reactions, dilute solutions, entropy constant. No math beyond calculus required. Problems. 160pp. 5⅜ x 8½. 60361-X Pa. $3.00

ELEMENTARY MECHANICS OF FLUIDS, Hunter Rouse. Classic undergraduate text widely considered to be far better than many later books. Ranges from fluid velocity and acceleration to role of compressibility in fluid motion. Numerous examples, questions, problems. 224 illustrations. 376pp. 5⅝ x 8¼. 63699-2 Pa. $5.00

UNCLE SILAS, J. Sheridan LeFanu. Victorian Gothic mystery novel, considered by many best of period, even better than Collins or Dickens. Wonderful psychological terror. Introduction by Frederick Shroyer. 436pp. 5⅜ x 8½.                                      21715-9 Pa. $6.00

JURGEN, James Branch Cabell. The great erotic fantasy of the 1920's that delighted thousands, shocked thousands more. Full final text, Lane edition with 13 plates by Frank Pape. 346pp. 5⅜ x 8½.
                                                    23507-6 Pa. $4.50

THE CLAVERINGS, Anthony Trollope. Major novel, chronicling aspects of British Victorian society, personalities. Reprint of Cornhill serialization, 16 plates by M. Edwards; first reprint of full text. Introduction by Norman Donaldson. 412pp. 5⅜ x 8½.                    23464-9 Pa. $5.00

KEPT IN THE DARK, Anthony Trollope. Unusual short novel about Victorian morality and abnormal psychology by the great English author. Probably the first American publication. Frontispiece by Sir John Millais. 92pp. 6½ x 9¼.                                     23609-9 Pa. $2.50

RALPH THE HEIR, Anthony Trollope. Forgotten tale of illegitimacy, inheritance. Master novel of Trollope's later years. Victorian country estates, clubs, Parliament, fox hunting, world of fully realized characters. Reprint of 1871 edition. 12 illustrations by F. A. Faser. 434pp. of text. 5⅜ x 8½.                                            23642-0 Pa. $5.00

YEKL and THE IMPORTED BRIDEGROOM AND OTHER STORIES OF THE NEW YORK GHETTO, Abraham Cahan. Film *Hester Street* based on *Yekl* (1896). Novel, other stories among first about Jewish immigrants of N.Y.'s East Side. Highly praised by W. D. Howells—Cahan "a new star of realism." New introduction by Bernard G. Richards. 240pp. 5⅜ x 8½.                                            22427-9 Pa. $3.50

THE HIGH PLACE, James Branch Cabell. Great fantasy writer's enchanting comedy of disenchantment set in 18th-century France. Considered by some critics to be even better than his famous *Jurgen.* 10 illustrations and numerous vignettes by noted fantasy artist Frank C. Pape. 320pp. 5⅜ x 8½.                                     23670-6 Pa. $4.00

ALICE'S ADVENTURES UNDER GROUND, Lewis Carroll. Facsimile of ms. Carroll gave Alice Liddell in 1864. Different in many ways from final Alice. Handlettered, illustrated by Carroll. Introduction by Martin Gardner. 128pp. 5⅜ x 8½.                             21482-6 Pa. $2.50

FAVORITE ANDREW LANG FAIRY TALE BOOKS IN MANY COLORS, Andrew Lang. The four Lang favorites in a boxed set—the complete *Red, Green, Yellow* and *Blue* Fairy Books. 164 stories; 439 illustrations by Lancelot Speed, Henry Ford and G. P. Jacomb Hood. Total of about 1500pp. 5⅜ x 8½.                      23407-X Boxed set, Pa. $15.95

AMERICAN ANTIQUE FURNITURE, Edgar G. Miller, Jr. The basic coverage of all American furniture before 1840: chapters per item chronologically cover all types of furniture, with more than 2100 photos. Total of 1106pp. 7⅞ x 10¾.   21599-7, 21600-4 Pa., Two-vol. set $17.90

ILLUSTRATED GUIDE TO SHAKER FURNITURE, Robert Meader. Director, Shaker Museum, Old Chatham, presents up-to-date coverage of all furniture and appurtenances, with much on local styles not available elsewhere. 235 photos. 146pp. 9 x 12.   22819-3 Pa. $6.00

ORIENTAL RUGS, ANTIQUE AND MODERN, Walter A. Hawley. Persia, Turkey, Caucasus, Central Asia, China, other traditions. Best general survey of all aspects: styles and periods, manufacture, uses, symbols and their interpretation, and identification. 96 illustrations, 11 in color. 320pp. 6⅛ x 9¼.   22366-3 Pa. $6.95

CHINESE POTTERY AND PORCELAIN, R. L. Hobson. Detailed descriptions and analyses by former Keeper of the Department of Oriental Antiquities and Ethnography at the British Museum. Covers hundreds of pieces from primitive times to 1915. Still the standard text for most periods. 136 plates, 40 in full color. Total of 750pp. 5⅝ x 8½.
23253-0 Pa. $10.00

THE WARES OF THE MING DYNASTY, R. L. Hobson. Foremost scholar examines and illustrates many varieties of Ming (1368-1644). Famous blue and white, polychrome, lesser-known styles and shapes. 117 illustrations, 9 full color, of outstanding pieces. Total of 263pp. 6⅛ x 9¼. (Available in U.S. only)   23652-8 Pa. $6.00